Understanding GMDSS

The Global Maritime Distress and Safety System

L. Tetley I. Eng., F.I.E.I.E.
Principal Lecturer in Communications Engineering

D. Calcutt M.Sc., C.Eng., M.I.E.E.
Senior Lecturer in Electronic Engineering

Edward Arnold
A member of the Hodder Headline Group
LONDON MELBOURNE AUCKLAND

© 1994 L. Tetley and D. Calcutt

First published in Great Britain 1994

British Library Cataloguing in Publication Data

Tetley, L.
  Understanding GMDSS: Global Maritime
  Distress and Safety System
  I. Title II. Calcutt, D.
  623.88

  ISBN 0-340-61042-5

Typeset in Great Britain by Computape (Pickering) Ltd. Printed
in Great Britain for Edward Arnold, a division of Hodder
Headline PLC, 338 Euston Road, London NW1 3BH, by
St. Edmundsbury Press, Bury St. Edmunds, Suffolk.

Preface

History has shown that changes are often made as a consequence of events. In addition, engineering changes are made as a consequence of the escalating development of technology. Both of these criteria have lead to a massive change in the methods by which distress alerting and communications are carried out on a global scale. The embryo of a new global communications system commenced operation in February 1992, and will continue to be developed during the transition period leading to February 1999 when the Global Maritime Distress and Safety System (GMDSS) becomes fully operational. The GMDSS is already extending the frontiers of mobile radio communications technology by building a complex highly reliable radionet to encompass the world, and take global mobile communications forward into the new millenium.

Not since the year 1912 and the tragic loss of the RMS Titanic has there been such a radical change in the field of maritime communications. Much has been written in retrospect about the unfortunate Titanic. A considerable amount of criticism has been levelled at the resultant rescue operation which, in fact, took place relatively efficiently within the guidelines valid at that time. However you view the tragedy, that singular maritime disaster instigated a necessary process of change which continues today. Maritime safety procedures are reviewed at regular meetings of the Safety of Life at Sea Convention under the guidance of the International Maritime Organization.

Clearly, a critical element within the safety package protecting all those who sail the oceans of the world is radio communications and its ability to alert other vessels and shore establishments of an impending disaster.

Radio communications within the GMDSS are composed of numerous elements forming a highly efficient radionet enclosing the entire surface of the world. Both terrestrial and satellite methods of communications are interlocked in order to produce an extremely reliable scheme whereby relatively inexperienced operators can use modern equipment to alert rescue authorities in the event of a vessel being in distress. However, whilst the use of satellite communications enables a shipboard operator to alert a shore-based rescue co-ordination centre rapidly, it should be remembered that a ship in distress requires immediate assistance which, in most cases, will be provided by shipping in the immediate area. Even considering the spectacular advances in technology, the human element still exists, and it is likely that distressed seamen will still be rescued by fellow seamen.

Gone are the days of the dedicated maritime radio officer who communicated by Morse code, often using a strange vocabulary. He/she has been replaced by the on-board operator whose secondary task it is to operate the radio communications equipment. Whilst mariners allocated the task of being shipboard radio operators are required to hold a Certificate of Competency, the skills-training period is very short compared with that of a dedicated radio officer, even compared with those who sailed on the ill-fated Titanic. However, the immense advances made in electronic equipment technology and system design have ensured that radio communications terminals can be operated very efficiently by virtually anyone.

This book has been written to provide the new radio operator and the electro-technical officer with all the knowledge required to understand fully the GMDSS and its related systems. There are three major sections in the book. Section one provides detailed information on the GMDSS itself. The complexity of the interlocking methods of communications, position fixing and distress work are carefully explained, along with specific equipment operational techniques. Section two looks at the massive impact of satellite communications and explains fully how, using satellites in general and the Inmarsat system in particular, distress alerting and communications via this medium have become relatively easy procedures. In addition, the operation of satellite Ship Earth Stations produced by major manufacturers is explained. The principles of terrestrial radio communications are considered in Section three, culminating in a description of some of the modern radio equipment available and its place within the GMDSS radionet. Finally, a detailed glossary and abbreviation section is provided which includes all those curious phrases, acronyms and buzzwords to be found in GMDSS literature and technical manuals.

This is a reference textbook which, when read in conjunction with technical manuals, will enable the reader to understand the technology employed in modern maritime communications systems. A large number of diagrams, photographs and illustrations have been included in order to make the text readily understandable.

The companion volume *Satellite Communications: Principles and Applications* has been written for use by maritime electro-technical officers, communications engineers and engineering students who require to analyse satellite communications system technology at a much greater depth.

L. Tetley
D. Calcutt
1994

Acknowledgements

A book of this complexity dedicated, as it is, to a brand new topic – the GMDSS – must inevitably owe much to the cooperation of various individuals, equipment manufacturers and organizations. To single out one or more organizations is perhaps invidious. However, our sincere thanks go to the parent and guardian of the GMDSS, the International Maritime Organization without whom the GMDSS would not be the excellent system that it is today

In many cases we have had no personal contact with individuals but despite this they gave freely of their time when information was requested.

We are extremely grateful for the assistance that the following companies and organizations gave during the writing of this book. We are particularly indebted to those organizations who permitted us to reproduce copyright material. Our sincere thanks go to the following:

Asea Brown Boveri (ABB) and ABB NERA AS.

BT Inmarsat Customer Services.

COSPAS-SARSAT Secretariat.

EB Communications (Great Britain) Ltd, supplier of ABB NERA GMDSS communications equipment.

The INMARSAT Organization.

The Inmarsat quarterly magazine *Ocean Voice*.

The International Maritime Organization (IMO).

Japan Radio Company Ltd (JRC).

Raytheon Marine Sales and Service Company, supplier of JRC GMDSS communications equipment.

The following figures are reproduced with the kind permission of the International Maritime Organization, London.

Figure 1.2, page 5. Figure 1.5, page 12. Figure 1.11, page 21. Figure 1.18, page 30. Figure 1.21, page 33. Figure 1.26, page 39. Figure 1.28, page 41. Figure 1.29, page 42.

The front cover picture depicting the GMDSS concept was derived from an original illustration produced by the IMO.

Last, but by no means least, both authors would like to thank their wives who provided constant support through the long hours spent pounding the keypad of the word processor.

In memory of Ray Cowhig

Contents

Section One

THE GLOBAL MARITIME DISTRESS AND SAFETY SYSTEM

1
The GMDSS radionet

1.1 Introduction

To many readers it may appear that the subject of global maritime communications is confusing, particularly when distress alerting and communications may be carried out by relatively inexperienced persons. The Global Maritime Distress and Safety System (GMDSS) has been developed to provide mariners with a global communications and locating network, the elements of which are capable of being operated by an individual with minimum communications knowledge and yet enable alerting and search and rescue (SAR) services to be reliably co-ordinated.

This system is of prime importance to all maritime personnel for it is likely that elements within the GMDSS will affect every individual in the future. Traditionally, the distress and alerting system, as defined by the International Convention for Safety of Life at Sea (SOLAS 74 and subsequent amendments) is constructed around the requirement that all vessels at sea keep a continuous radio listening watch on specific terrestrial radio frequencies. Readers may be aware of the three frequencies, 500 kHz in the medium frequency band (MF), 2182 kHz on medium frequency (MF) and channel 16 on very high frequency (VHF), all of which have been utilized for decades as international calling and distress channels. In the past, the radio equipment fitted on board a ship depended upon the type of vessel and the nature of the voyage undertaken, not, as in the case of the GMDSS, the geographical area in which the vessel is trading. Since, according to SOLAS 74, the minimum range which needs to be achieved by shipboard radio equipment is 150 nautical miles, assistance to a ship in distress would only be provided by other shipping close by. Whilst long-range radio communications have become easier within the GMDSS, the age-old tradition of seafarers rescuing seafarers in distress remains unchanged.

Traditionally, ships subject to the SOLAS 74 convention regulations and successive conventions relating to ships over 1600 grt, utilize two manually operated systems for distress alerting which are:

- Morse code telegraphy on 500 kHz MF,
- radio-telephony on 2182 kHz or 156.8 MHz (Channel 16) on VHF.

There are major disadvantages with both these systems. A highly trained Morse code operator is needed to handle the telegraphy alerting and communications traffic on 500 kHz. This fact requires ships to carry a dedicated specialized officer with all the subsequent ongoing costs. There is no place for the traditional Morse code operator in the GMDSS operation and consequently Morse telegraphy will cease to be used by ships operating under GMDSS regulations after 1999. The limited range of communications achieved using either MF or VHF radio-telephony can be tolerated and both frequency bands will still be used in the GMDSS, mainly during on-scene SAR operations. Because ships at sea have always had some difficulty in communicating over great

Fig. 1.1   An SAR operation in progress. (Courtesy *Ocean Voice*)

distances, satellite communications are increasingly being used on board international trading vessels. The rapid development of satellite communications and digital technology is having a major impact upon the availability of easy-to-use, reliable marine equipment. Satellite communications, using the excellent Inmarsat system, provide the mariner with virtually instantaneous access to a global radionet in the advent of distress. The system is also capable of vessel location and tracking—an invaluable function in the case where disaster rapidly overwhelms a ship. It should, however, be noted that whilst satellites provide instantaneous global communications they are by no means the only method available within the GMDSS radionet for distress alerting. Ships may use the more traditional method of terrestrial high-frequency communications.

1.2   The GMDSS system

During the GMDSS discussion phase the International Maritime Organization (IMO), the GMDSS organizing body, produced guidelines identifying system requirements.

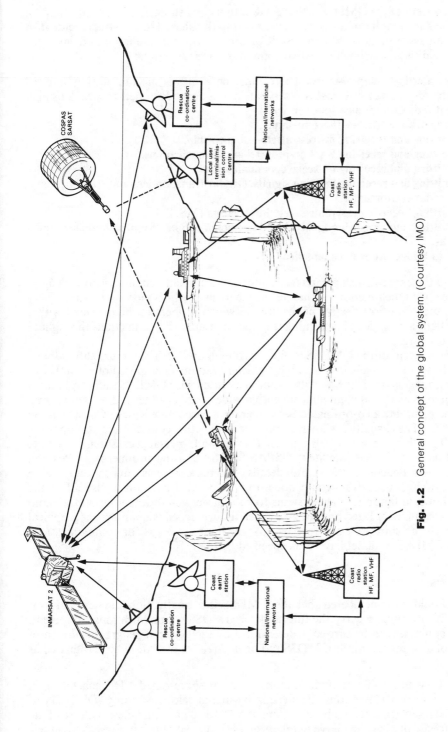

Fig. 1.2 General concept of the global system. (Courtesy IMO)

The basic concept of the GMDSS is that SAR authorities ashore, in addition to shipping in the immediate vicinity of a casualty, must be rapidly alerted to a distress incident so that they can assist in a co-ordinated SAR operation with the minimum of delay. This statement lead to nine defined principal communications functions.

1. **Distress alerting—ship-to-shore.** There must be a means of transmitting ship-to-shore distress alerts by at least two separate and independent methods, each using a different radiocommunications service.
2. **Receiving distress alerts—shore-to-ship.**
3. **Transmitting and receiving distress alerts—ship-to-ship.**
4. **Transmitting and receiving SAR co-ordination communications.**
5. **Transmitting and receiving on-scene communications.**
6. **Transmitting and receiving locating signals.** (Radar transponders and standard navigation radar equipment.)
7. **Transmitting/receiving maritime safety information (MSI).**
8. **Transmitting and receiving general radiocommunications from shore-based radio networks.**
9. **Bridge-to-bridge two-way communications.**

The IMO then specified ship carriage requirements for radiocommunications equipment to meet the nine functions in the specific sea areas in which the vessel trades. Every ship must, therefore, irrespective of the area in which it operates, be provided with a radio installation capable of performing the nine communications functions throughout its intended voyage.

The basic concept of the GMDSS is illustrated in figure 1.2. The picture shows that a ship in distress is effectively inside a highly efficient radionet composed of a number of interlocking subsystems. If the casualty is correctly fitted with GMDSS radio equipment it will be able to alert and communicate with a wide range of other radio stations and, through them, initiate a co-ordinated SAR operation based on a Rescue Co-ordination Centre (RCC). The casualty is shown communicating with a coastal radio station using VHF or MF, with two other vessels using VHF, with an Inmarsat geostationary satellite and broadcasting to a SAR satellite (COSPAS-SARSAT) using an Emergency Position Indicating Radio Beacon (EPIRB). Full details of all these systems can be found later in this book. A ship in distress would not normally be in a position to use all of the elements shown in figure 1.2. The systems to be utilized would depend upon the radio range of the equipment fitted on the ship, which in turn depends upon the geographical area in which the ship travels during its voyage. Four sea areas for communications within the GMDSS radionet have been specified by the IMO.

The sea areas

Whilst it should be remembered that the GMDSS is a totally global system it is not necessary for all ships to carry the full range of GMDSS communications equipment. The radio equipment to be carried is determined by the declared geographical area of operation of a vessel within the GMDSS radionet. Areas of operation are designated as follows.

● **Area A1**—within radio communications range of shore-based VHF coastal radio stations that provide continuous alerting by digital selective calling (DSC). Typically, the radio range would be approximately 20 to 30 nautical miles. Because of the huge expense of providing large numbers of VHF coastal radio stations around their coastline, many countries will not be establishing an Area A1. This means that ships must be fitted with radio equipment to satisfy Area A2 requirements when on home trading voyages.

- **Area A2**—within the radio range of shore-based MF coastal radio stations providing distress alerting using DSC. Typically, the radio range is 100 to 150 nautical miles.
- **Area A3**—within the coverage area of the Inmarsat geostationary satellites providing for continuous alerting. This is approximately the total surface area of the world excluding areas north and south of 75° latitude; or the total surface area of the world for those vessels which choose not to fit a satellite Ship Earth Station (SES) but use HF communications instead.
- **Area A4**—all other remaining areas of the world, namely the polar regions north and south of 75° latitude, outside the Inmarsat satellite coverage area.

Ship carriage requirements

Every vessel, over 300 grt, which chooses to operate subject to the GMDSS regulations must be provided with the following minimum fitting of radiocommunications equipment:

- a VHF radio installation providing communications on channels 6, 13 and 16 with the facilities for DSC alerting on channel 70;
- a receiver for continuous DSC watch on channel 70;
- two radar transponders (SART) transmitting in the 9 GHz maritime band;
- a NAVTEX receiver;
- a receiver for the reception of maritime safety information transmitted by Inmarsat's Enhanced Group Call (EGC) system if on voyages in sea areas of Inmarsat coverage where NAVTEX is not provided;
- satellite EPIRB capable of being manually or automatically activated to float free;
- two (three on ships over 500 grt) waterproof VHF hand-held transceivers for on-scene communications;
- MF communications on 2182 kHz (until February 1999).

Other equipment will be fitted depending upon the geographical areas through which the ship travels on its voyage. The equipment may be fitted in a dedicated communications position in the wheelhouse or in some remote position. However, GMDSS regulations require that the ability to initiate distress alerts must be provided at the position from which the vessel is normally navigated. This usually means that the MF/HF, VHF and SES equipment may be remotely operated from the wheelhouse. To simplify the initiation of a distress alert when a number of radio systems are fitted, a distress message controller (DMC) may also be fitted at each of the navigation and manœuvring positions. The DMC, as produced by JRC, using simple push button operation, initiates distress alerts on the three communications systems and produces audible and visual alarms on receipt of a distress alert.

The chart in figure 1.3(a) details the radio communications systems required for shipping in the GMDSS designated areas of the world. Figure 1.3(b) shows the radio communications equipment fitted on a modern GMDSS vessel trading in all areas. The equipment shown is provided by the Japan Radio Company (JRC) the world's largest manufacturer of GMDSS equipment and, from left to right, comprises; a 50 W public address system, fax machine, Inmarsat-A SES with printer on top, Inmarsat-C SES data terminal, NBDP for DSC communications, MF/HF radio-telephone and VHF radio-telephone.

The prime concern of any radiocommunications operator when a vessel is in imminent danger is that of distress alerting. GMDSS regulations require that each vessel

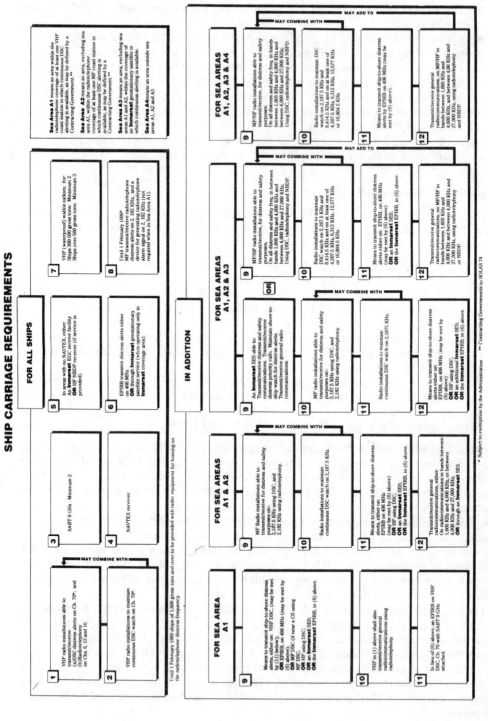

Fig. 1.3(a)   Ship carriage requirements. (Courtesy Inmarsat)

Fig. 1.3(b) A GMDSS installation on a modern vessel. (Courtesy JRC/Raytheon)

must be able to transmit a ship-to-shore distress alert by at least two totally independent means. The two pieces of equipment used for this purpose must be totally independent of each other including their power supplies. In most cases this means that the primary alert would be transmitted using the main communications terminal, possibly a satellite SES or a HF transceiver for ships in Area A3, with the secondary alerting, if required, carried out by using an EPIRB or another communications terminal. It is impossible to be specific because each vessel will be fitted with different equipment.

A GMDSS alert is usually initiated and acknowledged manually. Such an alert is easily initiated by using DSC on MF/HF, by pressing the 'red' button on an Inmarsat-A or B SES or by keyboard commands using an Inmarsat-C SES. The alert must be capable of being initiated from the position from which the vessel is normally navigated. However, if disaster overwhelms a vessel before an alert can be initiated, a float-free satellite EPIRB is automatically released and activated. This alert is received and decoded by the polar orbiting COSPAS-SARSAT satellites with a subsequent link into the GMDSS radionet.

The traditional international distress frequency is 2182 kHz on MF and has been retained for voice communications until February 1999. Channel 16 on VHF will continue to be used for voice communications following a distress alert sent by Digital Selective Calling channel 70. Channel 70 VHF and 2187.5 kHz are two new frequencies which are available immediately for distress alerting using DSC. In the interim period to 1999 it is necessary to have equipment which will maintain a listening radiowatch on a number of frequencies. In practice, a scanning receiver is used in which a number of spot frequencies are scanned along a number of bands.

Once the RCC, for a specific ocean region, has been advised of a distress alert it will use either terrestrial or satellite communications to relay the alert to other vessels in the area of the casualty. This implies shore-to-ship alerting using DSC on the VHF, MF or HF bands or an Enhanced Group Call (EGC) via the Inmarsat system.

After a confirmed distress alert has been received, SAR units are directed to the casualty using standard voice communications via either terrestrial or satellite means. 'Homing' as a method of locating the casualty is achieved by the use of a Search and Rescue Transponder (SART) radar beacon operating on the 9 GHz maritime band. Hence, other ships or SAR units are able to identify quickly a casualty when within radar range. In future, it is possible that a VHF radio direction finder will be a required fitting on SAR vessels in order to locate a casualty.

The GMDSS is therefore composed of a large number of interlocking elements all of which are considered in this section.

The transitional timetable for fitting GMDSS equipment on ships over 300 grt is as follows:

- existing ships may comply with GMDSS requirements from 1 February 1992;
- all ships must carry NAVTEX receivers and satellite EPIRBs from 1 August 1993;
- all new vessels must comply fully with GMDSS provisions from 1 February 1995;
- from 1 February 1999 all vessels must fully comply with the GMDSS.

1.3   Digital selective calling (DSC)

Digital selective calling forms a critical part of the terrestrial element of the GMDSS. Under international telecommunications regulations all transmitting stations must identify themselves and, consequently, each station is provided with a unique selective code. This may simply be the station name as in the name of a person or ship, or it may be the allocated group of call letters which were always used during Morse communications or, more commonly, it may be a group of call numbers. The sequential single-frequency code (SSFC) system uses audio frequencies in a four or five-digit code to identify marine coastal radio stations or ships. This outdated code is, however, soon

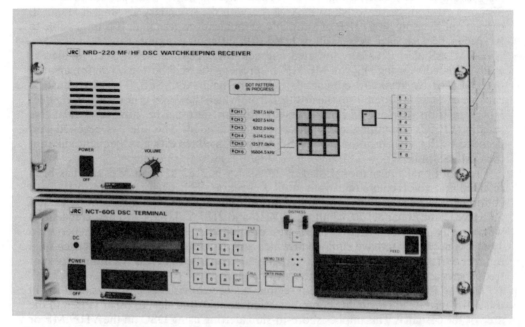

Fig. 1.4   Picture of DSC equipment. (Courtesy JRC/Raytheon)

to be replaced by the more reliable digital selective code calling system (DSC), in which the identities will be in decimal notation. Both ship and coast station identities known as the Maritime Mobile Service Identity (MMSI) number, will be coded in nine characters as shown below. All MMSIs include the three Maritime Identification Digits (MID) which identify the country of registration of the station.

(a) **Specific ship station identities.** $MIDX_4X_5X_6X_7X_8X_9$. The first three digits (MID) identify the country, followed by six digits to identify the ship station. As an example 257123456 indicates a ship registered in Norway as codes 257 and 258 are allocated to Norway.

(b) **Group ship station identities.** $0MIDX_5X_6X_7X_8X_9$. The first zero signifies a ship group call, the MID identifies the country assigning the call and the five digits identify the group of vessels being called.

(c) **Specific coast station identities.** $00MIDX_6X_7X_8X_9$. The first two zeros signify a coastal radio station. The remaining characters are as previously identified.

(d) **Group coast station identities.** $00MIDX_6X_7X_8X_9$. Characters as previously described.

Using the MMSI on DSC, collective calls may be made to all ships, or ships belonging to one company or trading in one area of the world. Selective calling is, depending upon radiowave propagational characteristics, a very reliable way of automatically calling ships. Although the ship's call number is shown in decimal format, DSC uses a sequence of seven unit binary combinations to transmit the number. It is only by the use of binary codes that true digital communications and hence reliability are achieved. Whilst DSC calls are of prime importance for distress alerting and acknowledgement, the system is capable of handling other routine communications.

DSC specifications

- **Frequencies**, either scanned or manually selected:
VHF DSC	Channel 70.
MF DSC	2187.5 kHz.
HF DSC	4207.5; 6312; 8414.5; 12,577 and 16,504.5 kHz.
MF NBDP	2174.5 kHz.
HF NBDP	4177.5; 6268; 8376.5; 12,520; 16,695 kHz.
- **Class of emission:**
MF/HF DSC	F1B.
VHF	F2B.
VHF Voice	G3E.
2,182 kHz voice	H3E.
MF/HF voice	J3E.

Figure 1.5 illustrates the sequence of distress call, relay and acknowledgement information which is transmitted in a DSC call. In the distress mode all messages will produce a hard copy on the associated printer. A distress call is initiated simply by pressing the 'red' button or keying a specific selected code. An incoming distress call initiates the narrow band direct printer (NBDP) along with audible and visual alarms. The transmission speed of a DSC alert varies depending upon the frequency band used. On MF/HF it is relatively slow at 100 bauds (twice the speed of standard telex), and on VHF where greater usable bandwidth is possible, it rises to 1200 bauds. The baud is the standard unit for expressing the speed of digital transmission.

With all methods of digital transmission it is necessary to include some form of error correction coding in the transmission. This leads to repetition of some data causing a

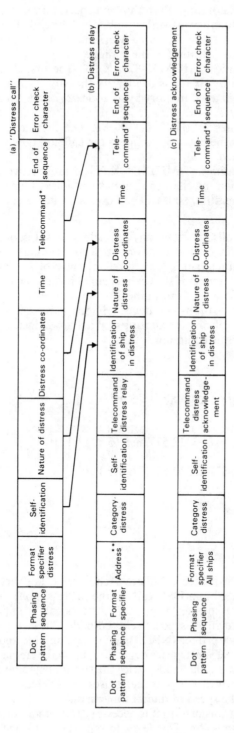

Fig. 1.5   Sequence of distress calls, distress relay call and distress acknowledgement. (Courtesy IMO)
* Type of subsequent communication (radiotelephony or teleprinter).
** Address is not included if the format specifier is "all ships".

high redundancy rate. However, the big advantage of the system is that suitable receiving apparatus can identify and correct errors caused in the transmission medium. A DSC sequence transmits each single character twice plus an overall message check at the end. A single call on MF/HF therefore varies between 6.2 and 7.2 seconds, whereas on the faster baud rate of VHF the same call would be transmitted between 0.45 and 0.63 seconds depending upon message content. In order to improve the chances of a DSC alert being received it is automatically transmitted for five consecutive attempts. Additionally, when a DSC alert is broadcast on MF or HF it is transmitted up to six times on any or all of the frequencies available (one on MF and five on HF).

Once the DSC broadcast has been activated the transmission format shown in figure 1.5 is sent automatically. The first two blocks of each message frame are essential in order to enable the receiving DSC unit to achieve synchronization. This is crucial in all data transmission systems to provide the receiver data decoder with the correct epoch information to enable message data to be received.

Distress alert data
- **Format specifier.** A distress code will automatically be sent in this frame.
- **Self-identification.** The unique nine digit number which identifies the vessel in distress.
- **Nature of distress.** This may be selected by the operator from one of nine codes, i.e. fire or explosion, flooding, collision etc. In the absence of a front panel input the system defaults to 'undesignated distress'.
- **Distress co-ordinates.** Automatically included from the interfaced satellite navigator memory or manually input. If no position is available the system defaults to 'no position'.
- **Time.** The time in UCT at which the distress co-ordinates were valid.
- **Telecommand.** Indicates whether subsequent distress information will be by radio-telephony or NBDP telegraphy. The system defaults to radio-telephony.

An 'end of sequence' code informs the receiver that the message has ended and to initiate an error checking operation to verify data.

Once a valid alert has been received and acknowledged by a RCC, a SAR operation will be initiated. On-scene communications are, by definition, short-range communications and will normally take place on VHF or MF between the casualty and other ships or aircraft in the area.

Locating a casualty may be done by using the 9 GHz SART transponder.

1.4 The GMDSS space segment

Although it is not necessary for ships to carry satellite equipment, satellite communications play a crucial role in the operation of the GMDSS. Suitably equipped vessels can transmit a distress alert and receive an acknowledgement virtually instantaneously irrespective of their geographical location. There are two satellite segments in the GMDSS each providing different services. They are:

- the Inmarsat global communications system providing instantaneous duplex communication based on geostationary satellite coverage;
- the COSPAS-SARSAT locating system based on polar orbiting satellites.

The latter system receives broadcast information only, no duplex communication is possible.

Inmarsat and the GMDSS

The description which follows outlines the Inmarsat system in relation to the GMDSS infrastructure and explains the crucial part which satellite communications play in the GMDSS radionet.

Section two of this book contains a description of the Inmarsat communications systems whilst a fully detailed technical description appears in the companion volume *Satellite Communications: Principles and Applications.*

Inmarsat is ideally placed to provide the GMDSS radiocommunications functions 1, 2, 4, 8 and 9 as specified by the IMO. The functions 3, 5 and 7 are of a short-range nature and will probably be carried out using VHF, although functions 5 and 7 could also be carried out using a satellite SES.

Instantaneous communications via satellite to a Coast Earth Station (CES) and then directly to a RCC provide the GMDSS function 1 requirements. Instantaneous access to a satellite is provided by the use of 'Priority 3 Distress' which is automatically included in a distress call made from an Inmarsat SES. Priority 3 establishes a satellite channel, or clears a channel if all satellite channels are engaged, directly to the RCC.

The GMDSS radiocommunications function 2 is provided by the RCC-to-ship alerting using Group calls to vessels within a designated sea area. SAR co-ordination communications can be provided between suitably equipped SESs to satisfy functions 4 and 8. Function 9 is ideally covered using the Inmarsat broadcasts of Maritime Safety Information (MSI) using the EGC service. This will include SafetyNET and FleetNET services.

Inmarsat-E EPIRBs provide for distress alerting by broadcasting to a geostationary satellite from an L-band EPIRB. Once activated the EPIRB provides distress alerting within 2 minutes directly to a shore station. The EPIRB transmits 20 alerts within a ten minute time frame, each alert contains the following information:

- the ship station ID;
- position information; and
- additional information to facilitate rescue.

Message data may be automatically interfaced with the EPIRB or input manually before release. The EPIRB must be mounted on the ship in such a position that the vessel's superstructure will not obstruct the L-band signals when operated *in situ*. It must also be float free from a mechanism which operates before reaching a depth of 4 m. As with all GMDSS equipment there must be a means of running a system diagnostic check at regular intervals without transmitting. A global coverage maritime Inmarsat-E map is shown in figure 1.6.

In addition to satisfying most of the IMO requirements within the GMDSS radionet, Inmarsat is able to provide other maritime safety features:

- automatic ship reporting or polling to enable shore authorities to know which vessels are in the area of a casualty,
- automatic transmission (ship-to-shore) of weather observations (OBS) to provide a detailed weather forecast for SAR units in the area of a casualty.

Inmarsat-A and distress

The Inmarsat Priority 3 Distress facility must only be used when there is grave and imminent danger to a vessel or a person.

When initiating a call, telephone or telex, and selecting a distress mode, some CESs

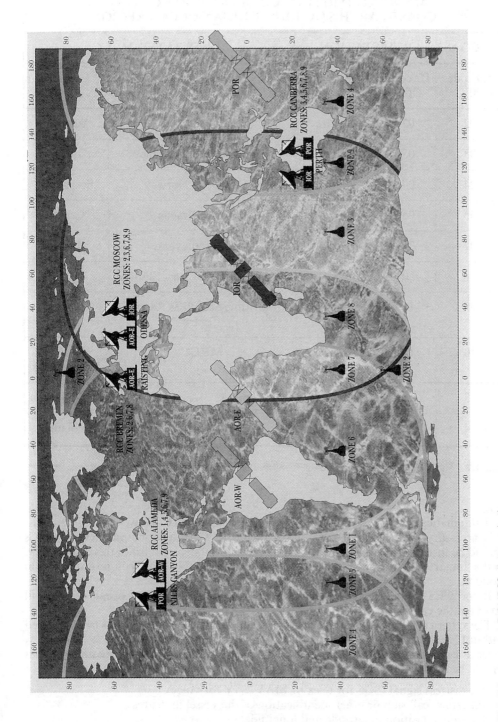

Fig. 1.6 Inmarsat-E/GMDSS distress alerting

ASSOCIATED RESCUE CO-ORDINATION CENTRES COAST EARTH STATIONS – INDIAN OCEAN REGION				
COAST EARTH STATION			ASSOCIATED MRCC	
ID code DEC	ID code OCT	CES Name	Authority	Location
02	02	Perth	MRCC Canberra	Australia
03	03	Yamaguchi	Maritime Safety Agency	Tokyo
04	04	Eik	RCC South Norway	Stavanger
05	05	Thermopylae	MRCC Piraeus	Greece
07	07	Odessa	v/o Moreplavanie	Moscow
08	10	Ata	Istanbul Radio Coast Station	Turkey
09	11	Beijing	Maritime Search & Rescue Centre	Beijing China
10	12	Burum	MRCC	Ijmuiden Netherlands
13	15	Jeddah	–	Saudi Arabia
14	16	Psary	RCC	Poland

Fig. 1.7   Table of associated RCCs for CES in IOR. (Courtesy Inmarsat)

will automatically route the call directly to an associated RCC. However, in other areas it is necessary to dial or key the number of the RCC. Telephone numbers of RCCs should be instantly available to the SES operator. Inmarsat publishes this information in the 'Inmarsat-A Maritime Users Guide' which is available from their headquarters in London.

If the CES is incorrectly selected, the NCS for the operating ocean region will intercept the call and redirect it to an RCC.

The system is very fast. Distress calls should be repeated if no reply has been received within 12 seconds.

Once contact has been established the SOLAS standard distress message format should be used as follows:

● the distress signal (SOS for telex or MAYDAY for voice);
● the name, call sign or other identification of the vessel in distress;
● the ship's position in latitude and longitude;

INMARSAT-B ACCESS REQUEST (DISTRESS) Frame S4									
BIT number									OCTET No
8	7	6	5	4	3	2	1		
Message type									1
SES ID									2
									3
									4
CES ID									5
Azimuth angle									6
Elevation angle									7
Service nature			Service type						8
Channel parameters									9
INIT/ repeat	TEST/ real	Spot-beam ID							10
CCITT CRC									11
									12

Fig. 1.8 S4 Access Request (Distress) frame. (Courtesy Inmarsat)

- the nature of the distress and the type of assistance required;
- any other information which may be of assistance to the RCC or SAR units.

Much of this information will be stored in memory and automatically transmitted when a distress alert is activated.

To avoid confusion it is essential that a distress message is formatted in this order. Figure 1.7 shows, as an example, the associated RCCs for CESs operating in the IOR.

Urgency and safety information can be obtained from some CESs by using standard Priority 0 and a specific two-digit code. An example of the codes used are:

- **Code 32: Medical advice.** The call should include the word MEDICO. Usually this call is automatically routed to a local hospital.
- **Code 38: Medical assistance.** Calls are routed to an RCC. The code must only be used when 'immediate' assistance is required.

● **Code 39: Maritime assistance.** Effectively the urgency call. To be used one level down from distress, when immediate assistance is required. For example, oil pollution, request for towage, etc.

Inmarsat-B and distress

Inmarsat-B uses the same Priority 3 distress access method as that used by Inmarsat-A and provides the same facilities for direct access to the GMDSS radionet.

Once Priority 3 access has been requested by the SES operator, the access procedure follows the same format as that described for Inmarsat-B operation in Chapter 7. Inmarsat-B signalling information is transmitted in frames of digital information. As an example the Access Request (Distress) frame is shown in figure 1.8. This frame contains all the necessary data to identify the casualty plus details of the azimuth and elevation angles of the ship's antenna with respect to the satellite. This information assists an RCC to locate the casualty if the vessel's position is corrupted.

If no spare channel is available to service the distress call, the CES will empty a communications channel to accommodate the call. Additionally, the NCS acts as back-up to the addressed CES should that CES fail to respond to a distress call.

The NCS always maintains a pair of global beam frequencies in reserve to ensure that a Priority 3 call will always gain access to the satellite.

The Inmarsat-B system contains a distress signal testing function whereby a SES initiates a test call by setting a 'test' flag in the distress access message frame S3. The CES responds to this with a recorded announcement followed by automatic clearing of the call.

Inmarsat-C and distress

As would be expected, maritime distress alerts and calls using the Inmarsat-C system share much commonality with the other two Inmarsat systems, the main difference being that a keyboard and display are used for communications instead of voice. An Inmarsat-C SES may be fitted with a 'red' distress button but it is more likely that the distress alert will be sent via the keyboard.

An Inmarsat-C display is usually menu driven and provides prompts to the operator requesting information. The standard information required may be as follows:

● SES IMN. SES Inmarsat number (ship's identity) entered automatically from software.
● CES identity. Select the nearest CES in your ocean region.
● Ship position. In latitude and longitude either interfaced directly from the satellite navigator or entered manually.
● Date and time of previous position update. Automatically or manually entered.
● Nature of distress. An operator selected option from a range of circumstances; collision, grounding, etc.
● Ship course and ship speed. Automatically or manually entered.

Figure 1.9 shows the ABB NERA AS Saturn-C display which uses pull down menus to set-up a distress call.

The next prompt is 'Do you want to send a **Real** Distress Alert?' If YES is selected the system instantly transmits the information to the selected CES. The resulting acknowledgement should come within a few seconds as shown in figure 1.10.

Urgency and safety information may be received on the Inmarsat-C system by using the EGC service, or information may be requested directly from shore by using a

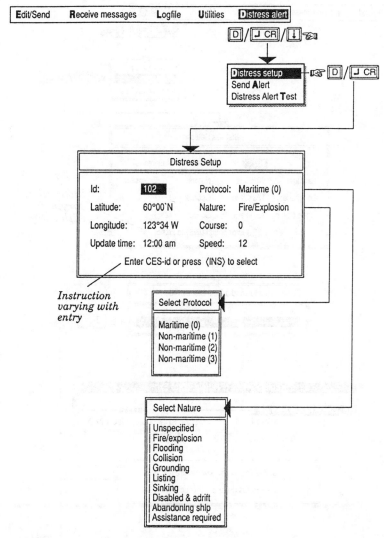

Fig. 1.9   Inmarsat-C distress set-up display. (Courtesy ABB NERA AS)

two-digit code, for example code 32 to request medical advice or code 41 to send a weather report. Land-based MESs are not permitted to use the maritime distress service although they may, optionally, send a land mobile alert.

The COSPAS-SARSAT alerting system

COSPAS-SARSAT is an international satellite system for Search and Rescue. The system consists of a constellation of satellites in polar orbit and a network of earth stations known as Local User Terminals (LUT). COSPAS-SARSAT provides distress alerting and location information, on a global scale, to appropriate RCCs.

The programme was created on 1 July 1988 when an inauguration agreement was signed between Canada, France, the former USSR and the USA.

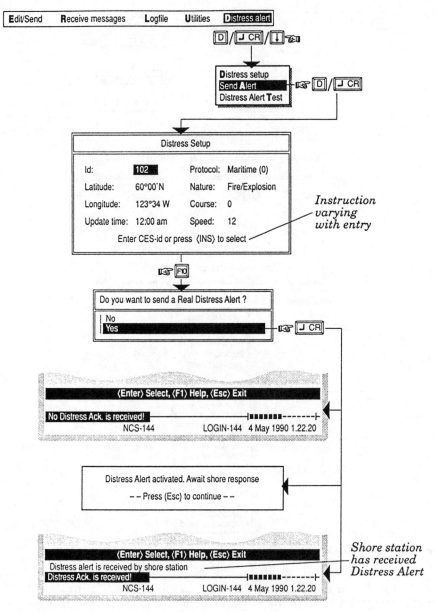

Fig. 1.10   Inmarsat-C sending distress alert. (Courtesy ABB NERA AS)

The COSPAS-SARSAT system uses a number of receive-only satellites in polar orbit that continuously monitor the earth's surface searching for emergency radio beacon signals transmitted on 121.5 or 406 MHz. Alert and location signals may be received from a maritime EPIRB, an emergency locator transmitter (ELT) or a personal locator beacon (PLB). The information is then downloaded to a LUT on earth and passed to SAR services via a Mission Control Centre (MCC) and the RCC. Figure 1.11. shows the basic system concept.

The space segment is based upon the NOAA range of satellites launched by the USA,

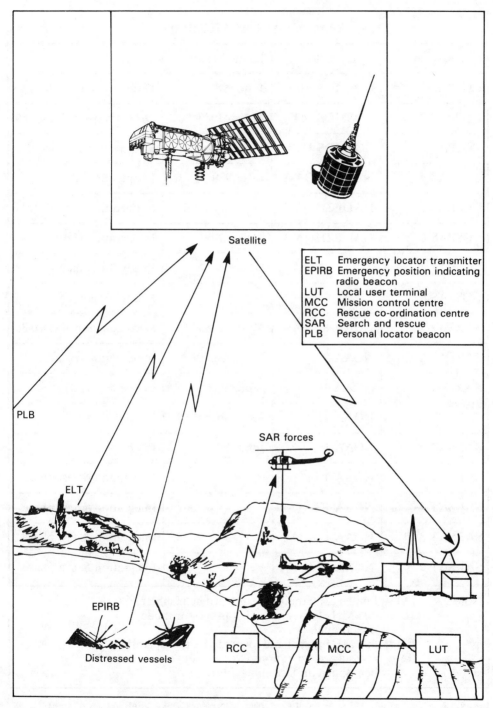

ELT	Emergency locator transmitter
EPIRB	Emergency position indicating radio beacon
LUT	Local user terminal
MCC	Mission control centre
RCC	Rescue co-ordination centre
SAR	Search and rescue
PLB	Personal locator beacon

Fig. 1.11 Basic concept COSPAS-SARSAT system. (Courtesy IMO)

SPACECRAFT AVAILABILITY			
C/S Payload	Spacecraft	Launch Date	Status
COSPAS-1	COSMOS 1383	June 1982	Decommissioned (Mar.88)
COSPAS-2	COSMOS 1447	March 1983	Decommissioned (Dec.89)
COSPAS-3	COSMOS 1574	June 1984	Decomissioned (Jun.90)
COSPAS-4	NADEZHDA-1	July 1989	In operation (A)
COSPAS-5	NADEZHDA-2	February 1990	In operation
COSPAS-6	NADEZHDA-3	March 1991	In operation (D)
COSPAS-7			Ready for launch
COSPAS-8			Ready for launch
SARSAT-1	NOAA-8	March 1983	Decommissioned (Dec.85)
SARSAT-2	NOAA-9	December 1984	In operation (B)
SARSAT-3	NOAA-10	September 1986	In operation (C)
SARSAT-4	NOAA-11	September 1988	In operation
SARSAT-5	NOAA-I	June 1992	In operation
SARSAT-6	NOAA-J		Integration & test phase
SARSAT-7	NOAA-K		Integration & test phase
SARSAT-8	NOAA-L		Integration & test phase
SARSAT-9	NOAA-M		Integration & test phase

Notes: (A) – limited availability in southern hemisphere.
           (B) – 406 MHz local and global processor mode
                operational.
           (C) – 406 MHz global and local modes not operating.
           (D) – C-6 started operations over the southern
                hemisphere in December 1991.

Fig. 1.12   COSPAS-SARSAT spacecraft availability. (Courtesy COSPAS-SARSAT Secretariat)

and the NADEZHDA satellites of the Russian Federation. The Russian Federation provides COSPAS satellites in near polar orbits at 1000 km altitude and the USA supplies NOAA satellites in sun-synchronous near-polar orbits at 850 km altitude. Each satellite carries an electronic package to receive emergency beacon signals transmitted on 121.5 MHz and 406 MHz. When viewed from the earth, each spacecraft is in view for between 5 and 15 minutes depending upon the maximum orbital elevation achieved in relation to the observer for a particular pass. Satellite orbital parameters are very similar to those of the Transit satellite navigation system described in the companion volume *Electronic Aids to Navigation (Position Fixing)*. Essentially, the earth footprint described on the surface by a single polar orbiting satellite will, due to earth rotation, cover the whole surface of the world. The orbital path of the satellite remains fixed while the earth rotates once every 24 hours beneath it. Clearly, more satellites placed in angular orbital planes to the first will search the earth's surface in a correspondingly shorter time. The COSPAS-SARSAT satellite availability in January 1993 is shown in figure 1.12.

Location techniques are very similar to those used in the Transit satellite navigation system. Upon receiving emergency beacon signals the satellite will measure the Doppler frequency shift, caused by the relative movement between the spacecraft and the beacon. This information and critical earth rotation timings are used by the LUT to calculate the position of a beacon on the surface of the earth. The use of Doppler location provides a simple method of locating a casualty. However, when used with the 121.5 MHz beacons, the process gives two positions for each beacon; the true position and its mirror image relative to the satellite ground track. This ambiguity is resolved by further calculation which again applies earth rotation parameters to the computation. Unlike the 121.5 MHz beacons, which were designed to be located by passing aircraft, the 406 MHz beacons are specifically designed for location by satellite and, consequently, the true position is determined on a single pass. A second satellite pass may be required to resolve the ambiguity present in the 121.5 MHz system. At 1000 km altitude, an orbital period is approximately 100 minutes, which may pose an unacceptable delay for the casualty if data from a second pass is required.

Typical alert sample messages are shown in figure 1.13. It should be noted that the 406 MHz beacon alert provides considerably more detail than the 121.5 MHz beacon message. The 406 MHz beacon message shows two predicted locations, 'A' with a 90% probability and 'B' with only a 10% probability, thus ambiguity is resolved on a single pass. In contrast, each location provided by the 121.5 MHz beacon has a 50% probability.

The spacecraft operate in two modes; 'local' mode and 'global coverage' mode. Both frequency systems operate in local mode, whereas the 406 MHz system also operates in global coverage mode.

Local mode repeater data system

In this mode a repeater on board the satellite relays signals directly to the ground to be received by an LUT in view of the satellite; or not as the case may be. This mode of operation provides quick alerts but only if a LUT is able to receive data at that instant. The world chart in figure 1.14 shows the coverage of 121.5 kHz beacons and LUTs.

406 MHz beacons are also able to operate in local mode. In this case the spacecraft recovers the data from the beacon and measures the Doppler shift. The information is time tagged and transferred, on 1544.5 MHz, to any LUT in view. It is also stored for downloading in the global coverage mode.

Typical 406 & 121.5 MHz Alert Messages sent out by an MCC

An example of a 406 MHz maritime beacon message;

  1. DISTRESS COSPAS/SARSAT ALERT CO2
  2. MSG NO 17034 REF NO 17009
  3. DETECTION TIME 04 APR 93 1708 UTC
  4. DETECTION FREQUENCY 406.026 MHz
  5. COUNTRY NORWAY
  6. USER CLASS MARITIME/IDENTIFICATION LEGA
  7. EMERGENCY CODE FIRE/EXPLOSION
  8. LOCATIONS   A. LAT 56 16.0N/LONG 006 48.4W PROB 90
                B. LAT 48 47.9N/LONG 049 37.0E PROB 10
  9. NEXT PASS   A. UNKNOWN
                B. UNKNOWN
10. REMARKS:

                A. HOMING SIGNAL, OTHER ACTIVATION
                MANUAL
                B. NIL
                C. NIL
                D. OPERATIONAL INFORMATION, TECHNICAL
                QUALITY EXCELLENT, TELEX NO 35721.
                VESSEL NAME
                ERIKA, GENERAL CARGO.

END OF MESSAGE

An example of a 121.5 MHz beacon message;

  1. DISTRESS COSPAS/SARSAT ALERT CO2
  2. MSG NO 17000 REF NO 16997
  3. DETECTION TIME 04 APR 93 1708 UTC
  4. DETECTION FREQUENCY 121.503 MHz
  5. COUNTRY NIL
  6. USER CLASS NIL
  7. EMERGENCY CODE NIL
  8. LOCATIONS   A. LAT 45 20.3N/LONG 075 51.2W PROB 50
                B. LAT 48 47.2N/LONG 087 59.0W PROB 50
  9. NEXT PASS   A. UNKNOWN
                B. UNKNOWN
10. REMARKS:
                A. NIL
                B. NIL
                C. NIL
                D. NIL

END OF MESSAGE

Fig. 1.13 Samples of alert messages sent by an MCC. (Courtesy COSPAS-SARSAT Secretariat)

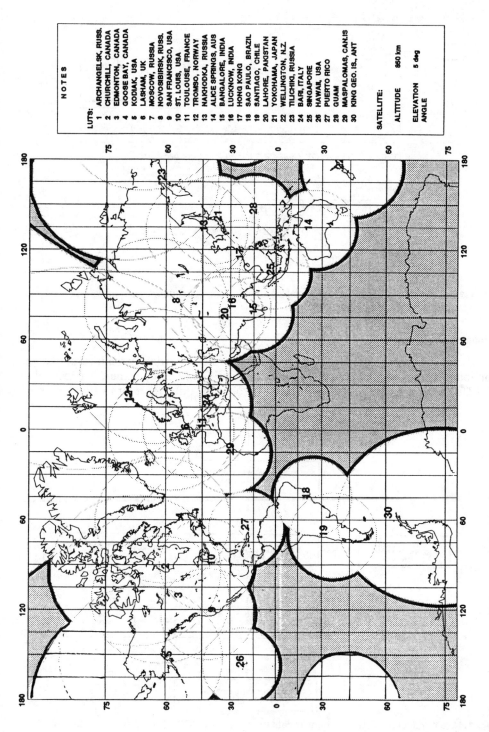

Fig. 1.14   Satellite visibility area of LUTs. (Courtesy COSPAS-SARSAT Secretariat)

NOTES

LUTS:

1	ARCHANGELSK, RUSS.
2	CHURCHILL, CANADA
3	EDMONTON, CANADA
4	GOOSE BAY, CANADA
5	KODIAK, USA
6	LASHAM, UK
7	MOSCOW, RUSSIA
8	NOVOSIBIRSK, RUSS.
9	SAN FRANCISCO, USA
10	ST. LOUIS, USA
11	TOULOUSE, FRANCE
12	TROMSO, NORWAY
13	NAKHODKA, RUSSIA
14	ALICE SPRINGS, AUS
15	BANGALORE, INDIA
16	LUCKNOW, INDIA
17	HONG KONG
18	SAO PAULO, BRAZIL
19	SANTIAGO, CHILE
20	LAHORE, PAKISTAN
21	YOKOHAMA, JAPAN
22	WELLINGTON, N.Z
23	TILICHIKI, RUSSIA
24	BARI, ITALY
25	SINGAPORE
26	HAWAII, USA
27	PUERTO RICO
28	GUAM
29	MASPALOMAS, CAN.IS
30	KING GEO. IS., ANT

SATELLITE:

ALTITUDE	850 km
ELEVATION ANGLE	5 deg

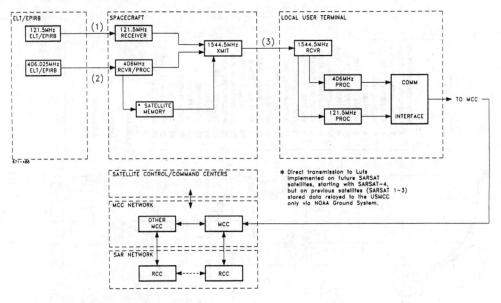

Fig. 1.15   System block diagram and satellite/ground segment interfaces. (Courtesy COSPAS-SARSAT Secretariat)

Global coverage mode

In this mode, beacon data which has been processed on-board the satellite is stored and continuously dumped down to LUTs when in view. Each beacon is therefore located by all LUTs causing a consequent reduction in processing time. In order to avoid unneces-

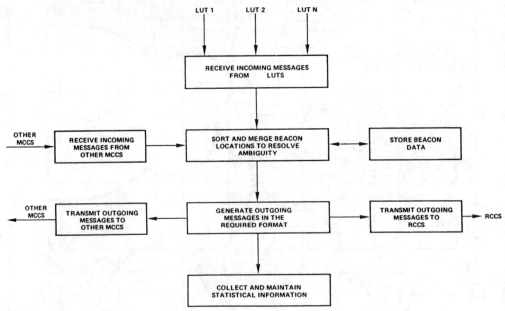

Fig. 1.16(a)   Functional diagram of alert data processing by MCCs. (Courtesy COSPAS-SARSAT Secretariat)

SAR 121.5 MHz Beacon Characteristics	
RF Specifications;	
Transmitted Power	50 – 100 mW PEP
Transmission Life	48 hours
Frequency	121.5 MHz +/− 6 kHz
Polarization	Linear
Modulation;	
Type	Amplitude
Depth	85%
Duty cycle	40%
Sweep rate	2 – 4 Hz
Modulation range	300 – 1600 Hz
The beacon may be used as a homing device by suitably equipped SAR units.	

Fig. 1.16(b)   SAR 121.5 MHz beacon characteristics

sary transmission of identical data from a number of LUTs, redundant data is sorted out in the MCC network and sent to the appropriate RCC.

A MCC is the control station for one or more LUTs operating within a country. Not only does the MCC process and rationalize beacon data received from LUTs, the station acts as a liaison with MCCs in other countries.

The MCC situated in the USA is the focal point for the co-ordination of SARSAT spacecraft operations. It distributes ephemeris data, processes time calibration data and forwards these results to other MCCs.

There are currently nine MCCs in operation, a further four are undergoing tests and four more are planned. This will eventually provide 17 MCCs operating in 17 countries.

There are also 17 LUTs in nine countries, six undergoing tests in five countries and seven more are planned.

It is evident therefore that the system is expanding and will continue to do so.

1.5   Emergency beacons

There are a number of emergency beacons, PLB, ELT, EPIRB, transmitting to satellites or aircraft, used in the GMDSS system. The beacons may be classified by frequency.

- 121.5 MHz—Transmitting to aircraft. Not currently accepted as a satellite EPIRB in

COPAS-SARSAT 406 MHz Beacon Characteristics (Taken from CCIR Recommendation 633)	
PARAMETER	VALUE
RF Signal:-	
Frequency stability; Short term (100 ms) Medium Term Mean Slope Residual freq. var.	2×10^{-9} $1 - 10^{-9}$/min $1 - 10^{-9}$
Power output.	5W +/− 2dB into 50 ohm load with a VSWR less than, or equal to, 1.25:1
Spurious emissions.	In-band mask defined for 406.0 – 406.1 MHz
Data encoding.	Bi-phase L
Carrier frequency.	406.025 +/− 0.005 MHz
Modulation.	Phase modulation of +/− 1.1 (+/− 0.1) radians peak
Modulation rise and fall time 150 +/− 100 microseconds.	
Digital Message:-	
Repetition rate.	50 sec. +/− 5%
Transmission time.	440 ms +/− 1% (short message) 520 ms +/− 1% (long message)
CW Preamble.	160 ms +/− 1%
Digital message.	280 ms +/− 1% (short message) 360 ms +/− 1% (long message)
Bit rate.	400 bps +/− 1%
Bit sync.	All 'ones' (15 ones)
Frame sync.	000101111
Continuous transmission: failure mode.	Transmission not to exceed 45s.
Temperature Range:- Minimum acceptable. Thermal shock.	−20°C to 55°C with long term temperature gradient of 5°C/hr. 30°C temperature difference with degraded performance for 15 minutes.
Minimum Operating Life Time	24 hrs. (COSPAS-SARSAT) baseline 48 hrs. (GMDSS)

Fig. 1.17   COSPAS-SARSAT 406 MHz beacon specifications.(Courtesy COSPAS-SARSAT Secretariat)

the GMDSS due to unavailability of global coverage. However, it is expected that the beacon will be used in future as an aircraft homing device by a liferaft.

- 243 MHz—These beacons are not part of the GMDSS.
- 406 MHz—COSPAS-SARSAT beacons transmitting in local or global mode to satellites.
  May be used as homing beacons by suitably equipped SAR units.
- 1.6 GHz L-band—These maritime EPIRBs provide instant access to the Inmarsat geostationary satellite network.
  The L-band unit is a sophisticated float-free EPIRB, known as Inmarsat-E, which is interfaced with ship sensors to provide SAR data in the uplink message.

1.6   Search and rescue (SAR)

A distress alert reaches an RCC, for a designated area, via terrestrial or satellite communications. Initial communications with the distress ship or aircraft will be made by the reciprocal link if the casualty is able to communicate. If the distress alert has been received via an ELT or EPIRB its position will be determined and the information passed to the nearest RCC to the casualty. It is not always easy to identify which RCC is in a position to offer immediate assistance. If the alert is received from an SES via a CES for instance, the CES is unlikely to be geographically close to the casualty. The CES will now alert an RCC, which again may not be the best situated to offer assistance. Figure 1.19 shows how this situation is resolved.

It is likely that a ship finding itself in distress may be a considerable distance away from the coastline of any country which may have an RCC—not all have!—and consequently the only hope for rescue lies with other vessels in the vicinity. Traditionally, this has always been the case where sailors rescue sailors. Within the GMDSS radionet it is possible, using DSC and EGC for an RCC to contact vessels in the area of a casualty, assuming of course that regular traffic movement reports have previously been sent.

If a distress call or alert is routed to a marine RCC (MRCC) the station will move quickly into action. The first function is to estimate the 'degree of uncertainty' of the distress position in order to determine the extent of the area to be searched. The next stage is to create an outline plan of the proposed operation and communicate it to all the SAR units likely to be involved. SAR units may be specific boats or aircraft designed for the task or may be one of a variety of vessels sailing in the area of the casualty.

Once SAR operations are commenced, the MRCC will notify the owner or agent of the vessel of the actions taken. Adjacent RCCs will also be informed as they may be able to offer assistance. On-scene communications between the casualty and SAR units will usually take place using VHF although it is possible to use Inmarsat SESs for this purpose.

Homing

SAR units may 'home' on the casualty, using radar if the distress vessel has activated a 9 GHz SART unit. A SART may be a manually activated or may be a float free unit which is automatically activated as the vessel sinks. An activated SART will respond to SAR ship or aircraft radar signals by generating a swept-frequency signal producing a line of 20 blips on a radar screen extending approximately 8 nautical miles *away* from the SART's position along its line of bearing. As the rescue vessel closes on the beacon the blip pattern changes to show a number of concentric circles as shown in figure 1.20.

A SART possesses an omnidirectional antenna which is essential if it is to respond to

TECHNICAL CHARACTERISTICS OF THE SATELLITE EPRIB: INMARSAT-E.	
Modulation.	Non-coherent binary frequency shift keying. (FSK)
Transmit Frequency.	1644.3 – 1644.5 MHz* 1645.5 – 1646.5 MHz**
Deviation.	-120 Hz (0); $+120$ Hz (1); tolerance $+/-$ 1%
Accuracy of clock frequency.	$+/- 2 \times 10^{-6}$/year
Frequency accuracy/stability; long-term accuracy (1 year) short-term stability.	$+/- 3 \times 10^{-6}$ (max) 1×10^{-8} for 1 minute
FSK switching time.	80% transmit power within 1.5 msec.
Transmit power.	nominal 1 W into antenna input $+$ 1 dB $-$ 3 dB
Antenna.	0 dBi nominal gain (shaped beam or herispherical)
Polarization.	Right-hand circular

Frame length;
  Data
  synchronization
  parity bits***
	100 bits
	20 bits
	40 bits

Code.	NRZ – L
Modulation rate.	32 bauds
Total transmission.	40 minutes
Number of transmissions.	4

Notes:
  *   Inmarsat first generation satellites.
  **   Inmarsat second generation satellites.
  *** Parity bits are used in the receiver for forward error correction of the distress message. (Up to four errors can be corrected)

Fig. 1.18 Technical characteristics of the satellite EPRIB. (Courtesy IMO)

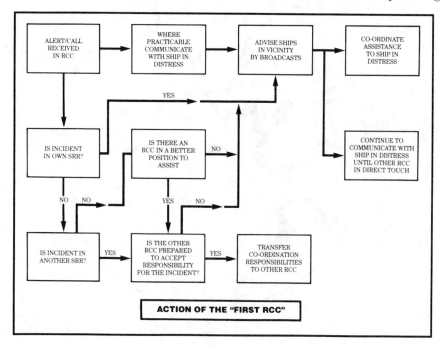

Fig. 1.19   Action of 'First RCC'. (Courtesy *Ocean Voice*)

SAR units approaching from any direction. GMDSS regulations require that a SART must operate correctly when interrogated by navigational radars with an antenna height of at least 15 m at a distance of up to at least 10 miles. Additionally, it must work if interrogated by aircraft radars with at least 10 kW PERP at an altitude of 2.5 km at a distance of 30 nautical miles. The technical characteristics of a SART approved by CCIR are given in figure 1.21.

1.7   Shore-to-ship alerting

Terrestrial shore-to-ship alerting is performed by voice using a vessel's unique individual callsign, or using its maritime mobile selective calling number (MMSI) on DSC on VHF, MF or HF bands. Alerting via a satellite is achieved using the Inmarsat EGC service with messages addressed to an individual vessel or a group of vessels in the SafetyNET service.

Enhanced group calling (EGC)

The EGC service has been established by Inmarsat to provide a fully automated service capable of addressing messages to individual vessels, pre-determined groups of ships or all ships in variable geographical areas. EGC alerts may be addressed to groups of ships designated by fleet, flag or geographical location.

A geographical area may be further defined as a standard weather forecast area, a NAVAREA, a rectangular area defined by latitude and longitude or a circular area around a maritime emergency.

The EGC facility meets SOLAS carriage requirements for sea areas A1 and A2 which are not served by the NAVTEX system.

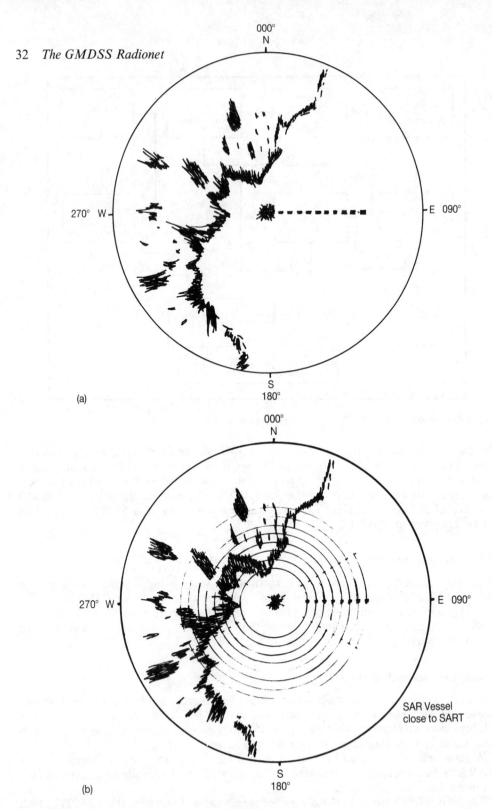

Fig. 1.20(a&b)   SART blips on a SAR radar screen

TECHNICAL CHARACTERISTICS OF SEARCH AND RESCUE RADAR TRANSPONDERS	
Transmitting frequency	9300–9500 MHz swept*
Polarization	Horizontal
Sweep rate	5 µs ± 0.5 µs
Form of sweep	Sawtooth, fast return < 1 µs
Pulse emission	100 µs nominal

Transmitting antenna
    Vertical beamwidth
    Azimuthal beamwidth
	
	< 25°
	Omnidirectional within ± 2 dB

EIRP	< 400 mW
Effective receiver sensitivity	Better than −50 dBm
Recovery time following excitation	Within 10 µs
Response delay	> 1.25 µs
Temperature range	−30°C to +65 deg. stowage −20°C to +55 deg. operation
*This frequency range should be extended to 9200 –9500 MHz, provided that the transponder also provides equivalent performance to that of a search and rescue transponder operating over the frequency band 9300–9500 MHz.	

Fig. 1.21   Technical characteristics of search and rescue radar transponders. (Courtesy IMO)

A ship fitted with an Inmarsat-C class 2 or 3 SES terminal, which includes an EGC receive facility, satisfies the SOLAS convention requirements for general communications in area A3.

Two EGC services are available: SafetyNET and FleetNET.

SafetyNET™

This service enables Information Providers who have been authorized by the IMO, within the auspices of the GMDSS, to distribute maritime safety information (MSI) from shore-to-ship as a safety service for mariners. The MSI service is an international system of radio broadcasts containing navigational information. Ship-board equipment automatically monitors the MSI frequencies and prints out information relevant to the ship.
Services include

- shore-to-ship distress alerts and urgent information from RCCs,
- meteorological warnings and forecasts compiled by National Weather Centres,
- navigational warnings and electronic chart correction data from Hydrographic offices,
- International Ice Patrol information for the North Atlantic Ocean.

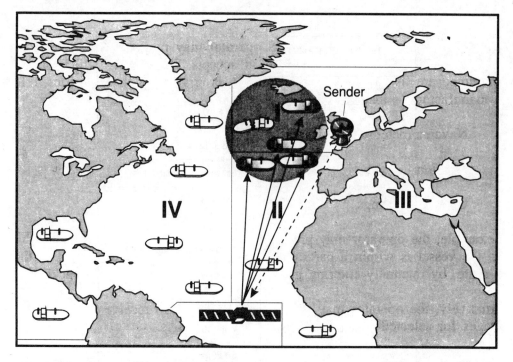

Fig. 1.22(a)   SafetyNET™ call to a circular area around an emergency. (Courtesy Inmarsat)

FleetNET™

The commercial FleetNET service enables authorized information providers, such as commercial organizations, governments and shipping companies, which have registered with a CES, to broadcast messages to selected groups of SESs. Services include:

● fleet or company broadcasts,
● commercial weather services,
● news broadcasts,
● stock market quotations,
● government broadcasts to all vessels sharing a country's registration.

EGC service codes

Once the equipment has been initialized to receive the NCS common signalling carrier for the ocean region in which it is situated, the operator can select some/all of the information listed by service codes. These codes provide an indication of the wide range of services available on the EGC network. See Figure 1.24.

Enhanced voice group call (EVGC)

Using a single voice channel on the Inmarsat-A or B service Inmarsat have provided an enhanced EGC service. EVGC provides for the reception of subscription services including the rebroadcasting of information services and radio programmes. In addition, slow scan facsimile can be accommodated within the restricted bandwidth of a voice channel to provide weather fax charts for any part of the world.

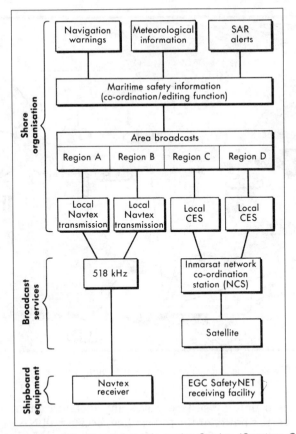

Fig. 1.22(b) The International Maritime Safety Information Service. (Courtesy *Ocean Voice*)

1.8 The NAVTEX service

The NAVTEX service forms an integral part of both the GMDSS and the world-wide navigational warning service (WWNWS) which is provided by Inmarsat using its EGC SafetyNET operation. These broadcast systems are designed to provide the navigator with up-to-date navigational warnings in English (the international GMDSS communications language).

Whilst NAVTEX and EGC SafetyNET message services require a number of different broadcasting arrangements, the two systems are similar. NAVTEX service areas are based on the IMO's 16 global NAVAREAS chart shown in figure 1.26.

Each NAVAREA is subdivided and covered by a number of transmitting stations. The subdivision of NAVAREA I is shown in figure 1.27(a).

In order that adjacent transmitters do not interfere with each other, a system of time division multiplexing (TDM) of the broadcast is used.

As an example, the TDM schedule for European stations is shown in figure 1.27(b). Similar station groupings occur in other parts of the world.

Service outline

NAVTEX is a terrestrial service using a single frequency of 518 kHz with frequency shift keying (FSK) and frequency modulation (FM) F1B designation. The medium frequency

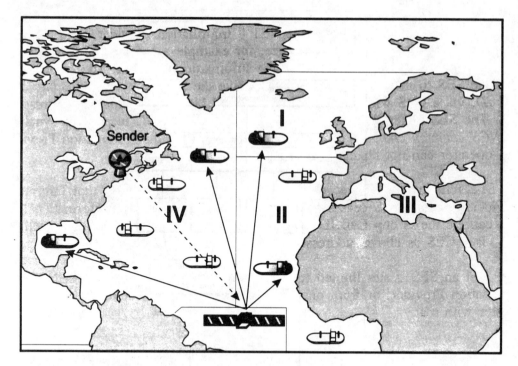

Fig. 1.23   FleetNET™ call to the ships of one company. (Courtesy Inmarsat)

518 kHz propagates mainly by surface waves and therefore the range is effectively determined by the carrier power at the transmitter. Because all NAVTEX transmitters world-wide use the same carrier frequency, the range must be strictly controlled. If two neighbouring transmissions were received by a single NAVTEX unit severe fading and signal degradation would result with a consequent loss of data. An additional safeguard against this occurrence is the use of TDM techniques on the carrier frequency. The simple organizational matrix used for worldwide NAVTEX TDM transmission is shown in figure 1.28(a).

Each NAVAREA is broken down into four groups, containing six transmitters each with a ten minute time allocation slot occurring every four hours. It should be noted that the matrix is designed for broadcasts of routine navigational information and that a considerable amount of data can be transmitted in ten minutes even at the relatively slow rate of 100 bauds. Distress and vital safety warnings are transmitted upon receipt during the empty slot periods.

Signalling codes

A dedicated NAVTEX receiver is used to receive incoming messages and print data on a Narrow Band Direct Printer (NBDP). The receiver possesses the ability to select messages to be printed according to a predetermined code. This code is designated B_1, B_2, B_3, B_4.

● Code B_1. This character is the transmitter identification character which may be used in the receiver to identify specific transmitters. Also, in order to prevent erroneous

ENHANCED GROUP CALLING SERVICE CODES	
Code	Meaning
00	All ships call
02	Group call
04	Met. warnings to rectangular areas
11	Inmarsat system message
13	NAVTEX re-broadcast
14	Shore-to-ship distress alert
22	WMO met. forecasts
23	EGC system messages
24	Met. warnings to circular areas
31	NAVAREA warnings
33	Download group identity
34	WMO met. forecasts to rectangular areas
42	WMO met. warnings
72	Chart correction services
Codes 00, 04, 14, 23, 24, 31, 33 & 42 cannot be rejected by a receiver.	

Fig. 1.24   Enhanced group calling service codes. (Courtesy Inmarsat)

reception by a receiver which is in a position to receive two transmissions using the same B_1 code letter, each B_1 code is allocated on the global pattern of NAVAREAS as shown in figure 1.26. Transmitters are given an alphabetical listing in sequence with no two transmitters, in surface radiowave range of each other, bearing the same alphabetical character.

The acceptable range for receiving NAVTEX transmissions is 400 nautical miles. This figure is based upon a transmitter carrier power of 1 kW and a receiver input sensitivity better than 1 μV with 10 dB signal-to-noise ratio.

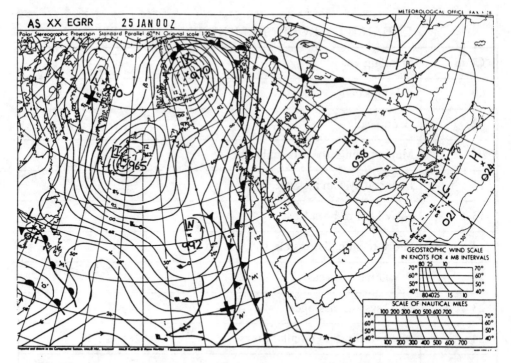

Fig. 1.25   Sample of a document which can be broadcast through EVGC. (Courtesy Inmarsat)

- Code B_2. This code is the subject indicator which identifies the different classes of message which are possible on transmission. The code B_2 is also used by the receiver to reject unwanted messages and to identify messages which cannot be rejected.
  The subject indicator characters for code B_2 are shown in figure 1.28(b).
- Codes B_3 and B_4. The two codes are used for message numbering, which will be in the range 01 to 99. The number 00 indicates that an extremely urgent message follows.

Message format

The technical format of a NAVTEX transmission frame is shown in figure 1.29. A 10 second synchronizing frame is followed by ZCZC which indicates the end of phasing. The 'B' code characters indicate coverage area, message type and numbering. Carriage return and line feed are included for NBDP control. The message follows and is concluded with NNNN. More printer control signals follow before the sequence is repeated.

Signal characteristics

FSK modulation is used to encode message data onto the 518 kHz NAVTEX carrier frequency. The FSK modulator shifts the carrier frequency either side of 518 kHz by ± 85 Hz. Thus, to encode a logic 0 the carrier is retarded to 517.915 kHz and for logic 1 it is advanced to 518.085 kHz conforming to CCIR recommendation 540. In the receiver, the 517.915 kHz signal is demodulated to an audio frequency of 1615 Hz representing logic 0, whilst the 518.085 kHz is demodulated to a logic 1 of 1785 Hz.

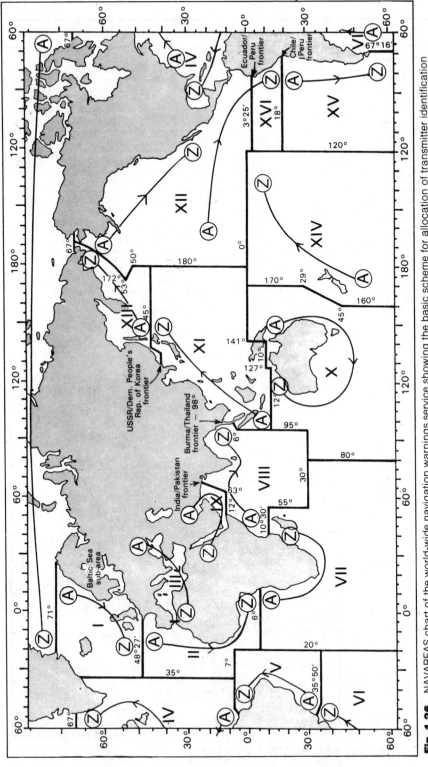

Fig. 1.26   NAVAREAS chart of the world-wide navigation warnings service showing the basic scheme for allocation of transmitter identification characters by IMO. (Courtesy IMO)

EUROPEAN TDM SCHEDULE FOR NAVTEX TRANSMISSIONS							
Code	Name	Times of transmission					
H	Harnosand	0000	0400	0800	1200	1600	2000
S	Niton	0018	0418	0818	1218	1618	2018
U	Tallinn	0030	0430	0830	1230	1630	2030
G	Cullercoats	0048	0448	0848	1248	1648	2048
F	Brest-le-Conquet	0118	0518	0918	1318	1718	2118
O	Portpatrick	0130	0530	0930	1330	1730	2130
L	Rogaland	0148	0548	0948	1348	1748	2148
T	Oostende	0248	0648	1048	1448	1848	2248
R	Reykjavik	0318	0718	1118	1518	1918	2318
J	Stockholm	0330	0730	1130	1530	1930	2330
P	Scheveningen	0348	0748	1148	1548	1948	2348
B	Bodo	0018	0418	0900	1218	1618	2100

Fig. 1.27(a)   European TDM schedule for NAVTEX transmissions. (Courtesy BT)

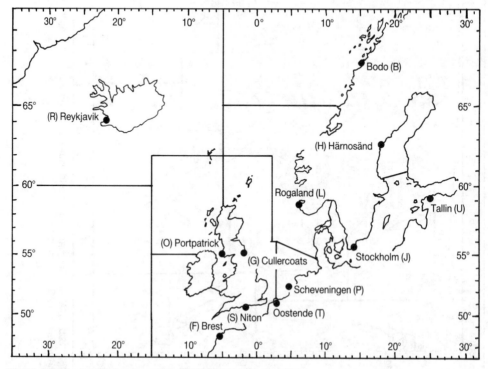

Fig. 1.27(b)   NAVTEX areas within NAVAREA I. (Courtesy BT)

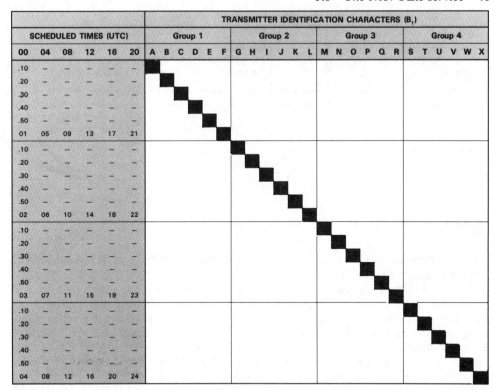

Fig. 1.28(a) NAVTEX global TDM transmission schedule. (Courtesy IMO)

NAVTEX SUBJECT INDICATOR CHARACTERS FOR CODE B_2	
Code	Meaning
A	Navigational warnings. *
B	Meteorological warnings. *
C	Ice reports.
D	SAR information. *
E	Meteorological forecasts.
F	Pilot service messages.
G	Decca messages.
H	Loran messages.
I	Omega messages.
J	SATNAV messages.
K	Other electronic NAVAID messages.
L	Navigational warnings additional to letter A. *
Z	No messages to transmit.
* Messages which cannot be rejected by the receiver.	

Fig. 1.28(b) NAVTEX subject indicator characters for code B2. (Courtesy IMO)

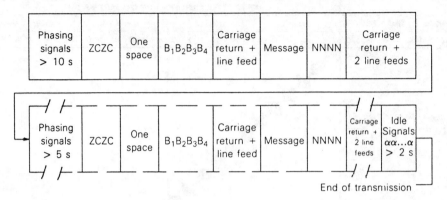

Fig. 1.29   Data format of NAVTEX transmissions. (Courtesy IMO)

Each alphanumeric character is serially encoded as a 7 data bit word (7-unit SITOR code) with a data rate of 100 bauds.

Figure 1.30 illustrates the complete NAVTEX coding standard which conforms to CCIR recommendation 476. There are however, only 35 possible combinations using this code and consequently each data string represents two possible characters. Data string 0010111 may represent T or 5. In order to eliminate this error each 7 bit data character is preceded by the letter or figure shift codes. In order to overcome errors caused by noise in the transmission path the NAVTEX system utilizes the same transmission protocol as that used by maritime radiotelex services—forward error correction (FEC). Each symbol is transmitted twice, the first time known as Dx (Direct) and the second known as Rx (repeat).

By reference to the coding standard chart, figure 1.30, the reader will observe that all the 7 bit codes possess four logic '1's and three logic '0's. This fact enables the demodulator to identify and correct a single bit error in the received signal. If either the Dx or the Rx words are corrupted the processor will print the other as the correct character. If both are corrupted an asterisk * is printed to indicate that the character is unreliable.

NAVTEX receiver operation

A NAVTEX receiver is a high gain dedicated microprocessor-controlled system which may be based on a tuned radio frequency (TRF) or a single superhet design.

The microprocessor is initially programmed to accept all transmissions on 518 kHz. Because of the receive sensitivity of around 1 μV this can mean receiving signals from NAVTEX transmitters in adjacent NAVAREAs. As this adjacent area information is of little importance and to save printer paper, it is usual to exclude, by using keypad commands, the messages which are not required.

The front panel arrangement of the Japan Radio Company JRC NCR–300A NAVTEX receiver shown in figure 1.31 illustrates how modern microprocessor-controlled equipment may be easily operated. A number of the control keys are self-explanatory. However, in some cases a combination of key commands is needed. As an example, it is a simple process to exclude unwanted transmissions. The PROG key used in conjunction with the A/M and E/D keys perform the task. Depressing the PROG key enables the use of the other keys. The A/M key alternately selects AREA or MESSAGE whilst the E/D key alternately selects the ENABLED or DISABLED

NAVTEX CODING STANDARD			
Data Input	HEX	Meaning	
		Letters	Figures
0001111	OF	carriage return	
0010111	17	T	5
0100111	27	8	?
100011	47	0	9
0011011	13	line feed	
0101011	28	no perforation	
1001011	48	H	
0110011	33	phasing signal q	
1010011	53	L	>
1100011	63	Z	+
0011101	1D	space	
0101101	2D	letter shift	
1001101	4D	N	,
0110101	35	E	3
1010101	55	R	4
1100101	65	D	$
0111001	39	U	7
1011001	59	I	8
1101001	69	S	
1110001	71	A	–
0011110	1E	V	=
0101110	2E	X	/
1001110	4E	M	.
0110110	36	figure shift	
1010110	56	G	@
1100110	66	phasing signal b	
0111010	3A	Q	1
1011010	5A	P	0
1101010	6A	Y	6
1110010	72	W	2
0111100	3C	K	(
1011100	5C	C	:
1101100	6C	F	%
1110100	74	J	BEL
1111000	78	phasing signal a	

Fig. 1.30   The NAVTEX coding standard

Fig. 1.31(a)   NAVTEX receiver front panel controls and message. (Courtesy JRC/Raytheon)

command. Thus, it is easy to enable or disable a NAVTEX area or message. The ALL key returns the equipment to the original condition. It should be remembered that certain messages cannot be rejected. Key command and display data charts are shown in figure 1.31(b).

If the STATE key is pressed the status of the receiver programming will be printed. A sample print out is shown in figure 1.32.

A typical NAVTEX transmission consists of a phasing sequence, a preamble, the message and then the 'end' signal.

For example:

> ZCZC SB03
> Nitonradio
> Dover wight SW winds expected storm
> force ten imminent.
> NNNN

ZCZC = Phasing sequence
S   = Nitonradio
B   = Category of message.
03   = Message number
NNNN = End.

Key	Meaning	Function
POWER	Power on	Turns the main power ON. Turns the main power OFF, when the OFF key is pressed continiously.
OFF	Power off `	Turns the main power OFF by using POWER key while OFF key is pressed continiously
TEST	Test mode	Performs a selftest.
FEED	Paper feed	Advances the printing paper.
MONI	Monitor	Turns the sound output of receiving signal from loud-speaker ON/OFF.
ILLUM	Illumination control	Controls the lightings of liquid Crystal Display (LCD) module, switch module and printing module.
A/M	Area/Message	Changes selection of AREA and MESSAGE alternately by pressing PROG key and holding it.
E/D	Enabled/ Disabled	Changes selection of ENABLED and DISABLED alternately by pressing PROG key and holding it.
PROG	Program	Makes A/M, E/D, ALL, AL OFF and SAVE keys effective by pressing this key first and holding it.
ALL	Selected all	Releases the rejection of AREA and MESSAGE by pressing the PROG key and holding it.
AL OFF	Alarm off	Stops an alert, without PROG key. Prohibits the alarm for message categories A and B OFF by pressing PROG key and holding it.
STATE	print-out status	Prints out the programmed status
SAVE	Store received messages	Prohibits printing of received message until pressing this key again. (Optional function)

Fig. 1.31(b)   NAVTEX receiver control panel and display functions. (Courtesy JRC/Raytheon)

A NAVTEX receiver system

The JRC NCR–300A NAVTEX receiver is similar to most receivers of this type. The omni-directional antenna may be passive or active and consequently a number of input stages are used to couple the antenna via protection diodes to the bandpass filter T1/T3. TR1 is a FET RF amplifier to provide gain before the signal is further filtered in FL1, the crystal bandpass filter. FL1 possesses a bandwidth of 500 Hz. The RF signal is now mixed in IC1 with a locally generated signal of 516.3 kHz to produce the baseband SITOR signal of 1700 Hz ± 85 Hz. This audible 1700 Hz provides both signal monitoring and the signal input to IC2 the FSK demodulator. IC2 is a standard PLL FSK demodulator which produces a logic 0 output when the signal input is 1785 Hz and a logic 1 when the input frequency swings to 1615 Hz. The data thus produced is fed to

Key	Meaning	Function
△	Move up	Move the symbol alphabet indicated on the LCD up through the alphabet.
▽	Move down	Move the symbol alphabet indicated on the LCD down through the alphabet.

Display	Function
(symbol)	Indicates the selected letter (B1 or B2), or indicates of the preamble being received during message receiving period.
AREA	Indicates that **AREA** on the LCD is being selected.
ENABLED DISABLED	Indicates whether the letter now being indicated is selected or rejected. **DISABLED** means rejected.
MESSAGE	Indicates that **MESSAGE** on the LCD is being selected.
ALARM DIS	(Alarm disabled) Indicates that the audio alarm for receiving message category A or B is prohibited.
RCV	Indicates that the equipment has been phased-in and is in receiving condition.
ALARM and MSG (Flickering)	Indicates that the message category A, B or D has been received.
ALARM and P-OUT (Flickering)	Indication that the printing paper have been running out.
SAVE	Indicates with lighting-up that the equipment has been under the SAVE mode, or indicates with flickering that the received messages are being stored in SAVE memory. (Optional function)

Fig. 1.31(b)   (cont.)

EXAMPLE :

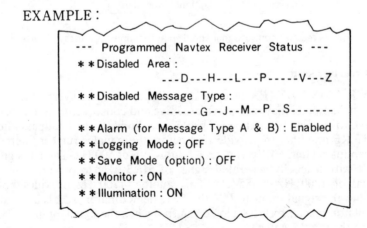

```
--- Programmed Navtex Receiver Status ---
* *Disabled Area :
            ---D---H---L---P-----V---Z
* *Disabled Message Type :
            ------G--J--M--P--S-------
* *Alarm (for Message Type A & B) : Enabled
* *Logging  Mode : OFF
* *Save  Mode (option) : OFF
* *Monitor : ON
* *Illumination : ON
```

Fig. 1.32   A NAVTEX receiver status display. (Courtesy JRC/Raytheon)

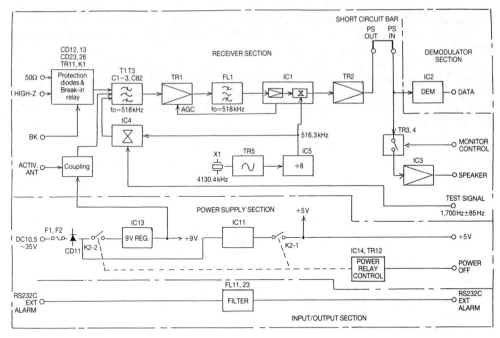

Fig. 1.33 A NAVTEX receiver block system diagram. (Courtesy JRC/Raytheon)

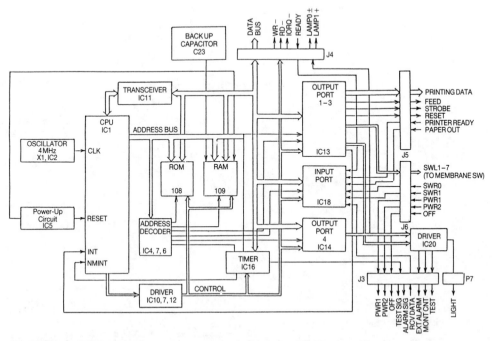

Fig. 1.34 A NAVTEX receiver processor board. (Courtesy JRC/Raytheon)

the microprocessor. During the self-test sequence a 1700 Hz ± 85 Hz signal is generated by the processor and mixed in IC4 with the 516.3 kHz oscillator signal to provide an input to the receiver board for testing.

Received data is input to the microprocessor via port IC18. IC1 is the ubiquitous Z80A CPU which controls the operation under the command of software held in IC8 the ROM. Processed printer data is output via IC13 to the printer. Other outputs from IC13 include printer control signals, keyboard scan signals and FSK PLL demodulator control signals.

2
Equipment maintenance requirements

2.1 Introduction

Obviously, it is of prime importance that the GMDSS communications equipment be maintained in proper working order.

In the past, most vessels have carried a radio officer whose job was to maintain all the ship's electronic systems in good working order. However, under GMDSS regulations there is no requirement for such an officer to be carried.

The IMO has declared that the GMDSS communications equipment shall be maintained in good readiness for use by employing one or more of the following procedures, not listed in any priority order:

1. the duplication of the GMDSS equipment,
2. shore-based maintenance,
3. at-sea electronic maintenance.

For ships trading only in areas A1 or A2, one of these options only need be chosen.

For ships trading in areas A3 or A4, any two of the options may be selected.

It should be remembered that the three options apply only to the regulatory GMDSS equipment and, consequently, it is only that equipment which needs to be duplicated. As most internationally trading ships carry an astounding array of other electronic equipment it is likely that responsible ship-owners will opt for the carriage of a qualified electronic service engineer, called an electro-technical officer. His/her duties include the maintenance of all electronic systems on board, which of course includes the GMDSS equipment.

However, it is possible for shipping companies to employ options 1 and 2, in which case no electro-technical officer needs to be carried.

Electronic equipment is vastly more reliable than was the case a decade ago. Even in a rugged seagoing environment, modern electronic communications equipment can be expected to continue to operate for periods in excess of 25,000 hours MTBF (mean time between failures). This equates to a period of continuous operation of almost 3 years without breaking down. A fact no doubt considered when the decision not to carry an electronics engineer on board ocean going ships was made.

Equipment intended for operation within the GMDSS radionet is designed so that main units can be replaced easily without the need for extensive circuit readjustment or recalibration. Adequate tools must be carried on board to enable the equipment to be properly maintained.

GMDSS equipment self testing

All communications equipment to be used in the GMDSS radionet must be fitted with some form of self-testing system in order that an operator may verify the equipment performance. In practice, different levels of self-test have been included and are noted, where appropriate, in individual sections of this book. As an example of a self-testing procedure the ABB NERA AS Saturn-C SES is shown in figure 2.1.

During a switch-on or restart procedure, the Electronics Unit performs function tests on the various electronic boards, cables and supplies. A combination of the blink rates of the five LEDs on the unit indicate the probable cause of any failure detected.

Power supplies

GMDSS equipment will operate from the ship's main and emergency sources of energy. However, a reserve source of energy must be provided as the sole supply to the GMDSS installation in the event of failure of the ship's systems. The reserve energy source, which will usually be batteries, must be capable of simultaneously operating the VHF radio installation plus the additional primary alerting radio installation appropriate for the sea area in which the ship is sailing. This may be a terrestrial MF/HF transceiver or a satellite SES.

The GMDSS reserve batteries must be totally independent of the ship's electrical system except for a battery charging arrangement.

The capacity of the reserve batteries must be such that the GMDSS equipment can be operated continuously for 1 hour on those ships which fully comply with the GMDSS regulations.

Battery capacity must be such that the total current drain is equal to the highest figure of all the radio equipment which can be connected simultaneously.

The radio equipment electrical current loading includes:

- the consumption of the VHF receiver,
- 20% of the consumption of the VHF transmitter,
- the total consumption of emergency lighting,
  and
- the consumption of an SES equipment on receive,
- 25% of the consumption of the SES when on transmitting maximum power.
  or
- the consumption of a MF/HF receiver,
- 33% of the consumption of the MF/HF transmitter when used for speech transmission at maximum power.

Lead acid batteries used for maritime electronic reserve energy applications tend to be 144 ampere hour capacity (AHC). That is, if the battery is discharged at a nominal 10 hour rate, in theory the battery would deliver a constant 14.4 A. In practice, it is not possible to discharge the battery totally and, consequently, this is a theoretical figure. However, when discharged at the 10 hour rate a fully charged battery would continue to deliver 14.4 A for between 6 and 7 hours. Of course, the battery could deliver 28.8 A for between 3 and 3.5 hours. The AHC rating is therefore very useful as an indication of the available energy contained in a battery.

Battery maintenance

No matter how reliable maritime equipment is made to be, it is often the battery supply which causes system failure. Not because battery technology is inadequate but because of insufficient or improper battery maintenance. In common with all things natural or man-made, batteries will not last forever. However, with periodic maintenance the secondary batteries which form the reserve energy supply for the GMDSS equipment

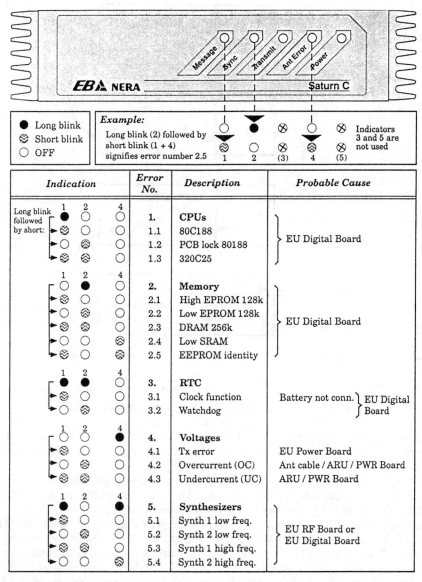

Fig. 2.1 An Inmarsat-C equipment display showing the results of a self-testing procedure. (Courtesy ABB NERA AS)

can be kept in prime condition. It is obviously critically important that the GMDSS emergency batteries are able to supply power when called upon to do so.

Batteries—the term used to describe a number of electro-chemical cells wired together in one container—come in two forms: primary batteries or cells, and secondary batteries.

Primary cells
These are cells which cannot be recharged and consequently their working life is limited by the duration of the chemical action within the cell. The duration of this action

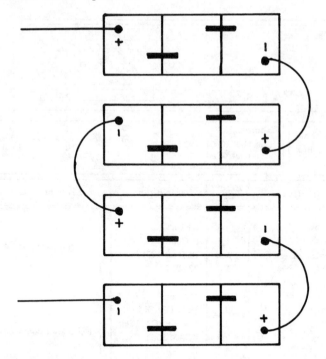

Fig. 2.2   Four batteries each containing three cells all joined in series to provide 24 V

depends upon the quantity of active chemicals which the cell contains. The larger the cell, the longer it will be able to deliver power to the equipment.

A primary cell used in a torch light is usually a Leclanche cell which provides a terminal voltage of 1.5 V. Hence, two, three or possible four are used in series to provide the correct voltage to operate the bulb. This elderly cell is not used in GMDSS equipment.

Modern primary cells are available in several types. They are classified using the name of the major chemical which they contain. There are currently three main types available; alkaline (1.5 V), mercury (1.4 V) and lithium (1.45 V).

A particularly important feature of all cells is the period known as 'shelf life'. This is the period of time which expires between manufacture and the time when the cell goes into service. Once a cell has been manufactured a local chemical action commences. This action is a form of self-destruct which, in time, reduces the effectiveness of the cell. In some older cells this local chemical action caused the cell container to be gradually destroyed and the leakage of corrosive substances damaged electronic equipment. With modern primary cells this is extremely unlikely to occur. However, the local action slowly reduces the quantity of energy stored within the cell. As an example, energy loss in primary cells due to this action is approximated as follows:

Leclanche cell—20% energy loss in 12 months,
alkaline cell—10% energy loss in 30 months,
mercury cell—20% energy loss in 5 years, and
lithium cell—2% energy loss in 5 years.

Primary cells used in the GMDSS equipment, EPIRBs for instance, tend to be lithium cells.

Secondary cells

Secondary cells are better suited to supplying the reserve power requirements of equipment operating within a rugged environment, the GMDSS. These cells are able to be discharged and recharged many times. In addition, they are robust and able to supply much greater amounts of energy on demand. Secondary cells are not used singly but will be grouped together in batteries providing a terminal voltage of 6 V or 12 V. Obviously they can be further grouped to provide multiples of these voltages. There would normally be, on board a ship, two banks of batteries each with a nominal potential difference (pd) of 24 V. Each 24 V bank of batteries would be composed of four 6 V batteries wired in series to produce a terminal voltage of 24 V. Each 6 V battery contains three 2 V cells. Two volts is the nominal terminal voltage of a lead acid cell. This voltage will change only slightly when the battery is being charged or discharged.

The basic principle of any electro-chemical cell is a positive plate containing one chemical, and a negative plate containing another, immersed in an electrolyte. With secondary batteries the electrolyte is a liquid (or jelly) and probably accessible for maintenance.

The lead acid secondary battery is commonly found in most high-energy situations although the nickel cadmium (NiCad) battery is also available.

Lead acid batteries

The main reasons for the continued use of lead acid batteries are:

(a) the energy density tends to be very large,
(b) they are able to work reliably over large temperature ranges,
(c) they are very robust and will stand some ill treatment,
(d) they tend to be cheap to purchase, and
(e) if well maintained they will reliably deliver energy for many years.

A lead acid battery on discharge produces electricity because of a chemical reaction which occurs between the active material on both the positive and negative plates and the electrolyte. In this type of cell the electrolyte is a solution of sulphuric acid and distilled water. Once the battery is discharged it may be recharged by connecting it to a battery charger. The charger causes a direct current to flow into the battery, reversing the chemical action to cause energy to be stored in the plates. During the discharge/charge cycle the 'specific gravity' of the electrolyte undergoes considerable change. Because the terminal voltage of a lead acid cell changes little, it is this change of specific gravity which provides a reliable indication of the state of charge of each cell.

The specific gravity (SG) of the electrolyte changes linearly as the cell is charged or discharged, as shown in figure 2.3. It is therefore a simple matter to estimate how much energy the battery contains. Depending upon temperature, the SG of a fully charged lead acid cell is approximately 1230 and the cell should not be permitted to discharge below a SG reading of 1180.

The specific gravity reading of a cell is easily checked using a float device called a hydrometer. It must be remembered that the electrolyte is dilute sulphuric acid which can cause serious injury if it comes into contact with eyes or skin. Always follow the simple safety and cleanliness rules when using a hydrometer to check the SG reading.

(a) Whilst wearing rubber gloves, goggles and protective clothing, remove the top cap of the cell to be checked.
(b) Hold the hydrometer over the open cell and squeeze the rubber bulb to expel the air. Never squeeze the bulb when the tip of the hydrometer is immersed in the electrolyte as this may cause the corrosive electrolyte to be sprayed upwards.

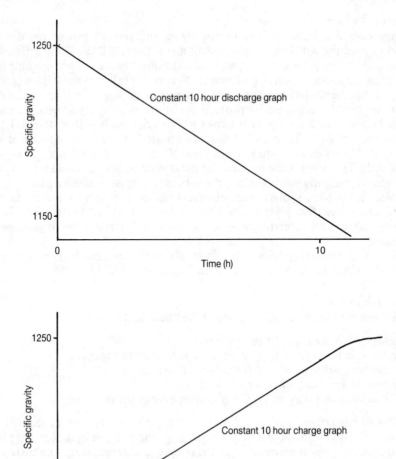

Fig. 2.3   Ten hour rate discharge and charge SG graphs for a lead acid cell

(c) With the tip of the hydrometer immersed in the electrolyte, slowly release the pressure of the bulb and draw sufficient liquid into the tube to enable the float to lift clear of the base.
(d) Without removing the hydrometer from the cell note the reading at the point on the float which is level with the liquid in the tube.
(e) Gently squeeze the bulb to return the electrolyte to the cell without causing splashing.
(f) Replace the top cap and, using a clean disposable cloth, clean and dry the top of the battery.

Whilst it is essential to use the SG reading as an indication of the charge state of the

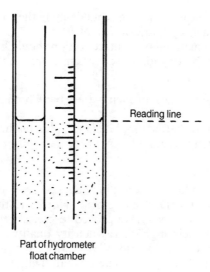

Reading line

Part of hydrometer
float chamber

Fig. 2.4 Reading the hydrometer

batteries, it is also possible to reach a quick estimate of that state by placing the batteries 'on-load'. The nominal terminal voltage of the battery on discharge will be 24 V. However, if the battery is used on-load in a discharged state, for instance to drive the MF/HF transmitter on full power, the terminal voltage will reduce. The extent of this reduction depends upon the discharge state of the battery bank. If the bank is almost fully discharged the terminal voltage, or potential difference (pd), is likely to drop to approximately 20 V. A fully charged bank of batteries used under the same load conditions would have a pd of approximately 23 to 23.5 V. Therefore, the greater the drop of pd when used on-load, the greater the discharge state of the batteries.

Similarly, when charging the bank of batteries, the greater the discharge state, the greater will be the direct current required to recharge the batteries. The greater the current applied to the resistance, which is the battery, the larger will be the pd, which could indicate in excess of 28 V when on charge.

In practice, two banks of reserve batteries are provided in order to maintain reserve energy supply. One bank will be used on discharge to drive the equipment required, whilst the other may be on charge to keep it in a fully charged condition. If both banks are switched in parallel, both will be discharged together and the total energy available will be the sum of that contained in the two banks.

Lead acid batteries should be regularly 'cycled' in order to maintain battery efficiency. Cycling requires that the batteries should be partially discharged and recharged at regular intervals. The provision of two banks of batteries enables cycling to be easily achieved without interruption of supplies to the equipment.

There are specific rules governing the design and location of the reserve battery locker. In practice, the locker will be located as high as practical on the vessel and as close as possible to the equipment which is to be supplied with energy. If a ship is sinking it is expected that the reserve batteries will continue to supply energy until seawater reaches them. Hence, the battery locker is as high as practicable on the ship. In addition, long cable runs between battery and equipment possess relatively large resistance and consequently will cause loss of voltage.

Lastly, the battery locker must be well ventilated. When charging lead acid cells, the inflammable gas hydrogen is released. If the gas builds-up in the locker an explosion will result if the gas is ignited by a stray spark or a smoker!

A detailed record of battery maintenance and condition should be kept. In addition to full details of the batteries, the battery logbook contains details of the charge state of the battery banks as follows:

(a) daily entries of the terminal voltage of each battery bank both off-load and on-load, and
(b) monthly entries concerning battery maintenance, SG readings per cell and full details of any charge which is required.

Nickel cadmium batteries

Nickel cadmium (NiCad) secondary batteries may also be used to provide the reserve source of power for the full GMDSS equipment. However, they tend to be very expensive and, in some situations, they have proved to be less reliable than their lead acid equivalents. Because of their high energy-density quality, NiCad batteries will normally be used to power the hand-held VHF transceiver. Size for size, NiCad batteries are able to store considerably more energy than their lead acid cousins, which makes them ideal for use in hand-held equipment.

The capacity of a NiCad battery is defined as the quantity of energy which the battery can deliver over a defined period of time: typically 1 hour in the United States and 1 to 5 hours in Europe. The life of a NiCad battery is dependent upon many factors:

(a) the quality of the construction,
(b) the operational environment,
(c) its suitability for the job,
(d) the design of the battery charger, and
(e) the number of duty cycles.

Most NiCad batteries would be expected to maintain their efficiency through approximately 1000 duty cycles: that is, one thousand charge/discharge cycles before the capacity of the battery falls to below 80% of the nominal figure. However, it is not advisable to expect the battery to reach its absolute limit when it is used to power GMDSS equipment. NiCad batteries will be changed before their efficiency starts to decline.

New, unused NiCad batteries may exhibit a temporary loss of capacity if they have been sitting on the shelf for over three months. They are 'lazy' because, if they are immediately charged and put straight into service, their efficiency will have dropped to about 40%. Lazy new NiCad batteries need to be cycled three times to improve their efficiency before being put into active service.

The electrolyte contained in NiCad batteries does not provide a reliable indication of the charge condition of a cell. Consequently, the terminal voltage, which changes slightly on discharge, may be used to indicate charge condition. For this reason terminal voltage is displayed on modern electronically controlled battery chargers.

The discharge voltage curve shows that the terminal voltage drops approximately linearly from the high of 1.35 V to the low of 1.1 V after 5 hours of continuous use. A NiCad battery is quoted as having a fixed duration charge time of say 5 hours as shown in figure 2.5, which must be strictly controlled to prevent internal damage to the cell. Charging is the key to obtaining maximum performance and life. Most NiCad batteries are able to accept a high level of current when charging. However, when the battery becomes fully charged, the charging current must be limited to the quoted maximum overcharge rate for the cell. This is usually $C/10$, or one tenth of the nominal charging

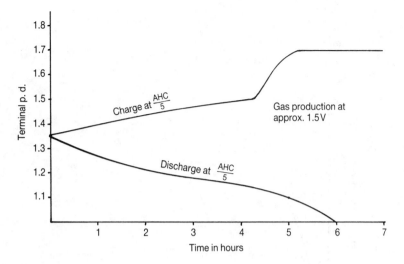

Fig. 2.5   Graph showing the terminal voltage changes of a NiCad cell over a five hour duty cycle

rate. A NiCad battery usually requires between 140% and 160% of its rated AHC in order to return it to a fully charged state. Modern electronically controlled battery chargers automatically sense the terminal voltage of the NiCad battery and adjust the charging rate accordingly, in order to achieve full charge without causing severe overcharging. The charger usually possesses a red light to indicate that the battery is correctly charging followed by a green light to indicate that the battery is ready for use.

As with lead acid batteries it is possible to 'trickle' charge NiCad batteries. A trickle charge is usually one tenth of the normal rate, causing the battery to require between 14 and 16 hours to become fully charged.

Good practices when using NiCad batteries
- Do follow the manufacturers' instructions correctly.
- Do keep all battery and charger contacts clean.
- Do store spare batteries in a cool location. Long-term storage at elevated temperatures causes high self-discharge and progressive deterioration.
- Do check the capacity, by the use of charge/discharge cycles, every three months.
- Do rigidly follow all instructions regarding battery charging.

Bad practices when handling NiCad batteries
- Do not short circuit the battery terminals. This will blow the internal fuse, if one is fitted. If no fuse is fitted the short circuit will destroy the battery.
- Do not repeat fast charging cycles in the hope of increasing battery capacity. Repeated charging will build-up heat within the case and may cause an explosion.
- Do not fast charge batteries that are below 5° C.
- Do not fast charge batteries which have been stored for 6 months or more. The initial charge should be a slow charge of about 14 to 16 hours.
- Do not dispose of batteries by dumping them over the side. Think of the environment. Remember batteries contain harmful chemicals. Dispose of them correctly at the next port.

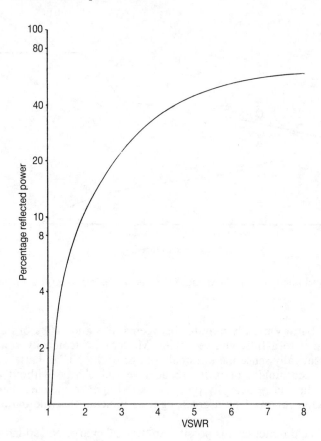

Fig. 2.6   Percentage reflected power against VSWR

Antenna maintenance

Antenna systems do not require constant maintenance but they do need to be inspected at intervals and particularly after the ship has been through storms and heavy weather.

Physical damage to antennae should be obvious. Less obvious though is the damage which is caused to the feeder by ingress of water. All feeders, or coaxial cables are connected to an antenna in a way which will exclude water. If the rubber caps, sleeves or containers covering the connection have been damaged they must be immediately replaced. The connection should be dried, and if water has entered the coaxial feeder cable this too should be replaced.

Dried sea salt or exhaust deposits from a ship's funnel may form a conductive surface on antenna insulators leading to a loss of transmitted radiated power. Insulators can be inspected visually or, during the hours of darkness, it may be possible to see the RF energy arcing across the insulator.

Another loss to radiation efficiency is caused by RF energy being lost from bad connections. This may be seen at night as a corona discharge into the atmosphere.

Mechanical vibration and regular movement can cause feeders to become detached from the antenna. It is essential therefore that the feeder coaxial cable is secured to the ship's superstructure at regular intervals.

Good practices for antenna maintenance
- Do check the antennae regularly, particularly after heavy weather.
- Do ensure that all insulators are clean. If the surfaces are contaminated, clean them with desalinated water. Do not use an abrasive cleaner.
- Do ensure that all weather proofing covers are in place and secure.
- Do ensure that all coaxial cables connecting the antennae with equipment are undamaged and securely fixed to the ship's superstructure.
- Do make sure that any 'earth' connection which is made by bonding the outer screen of a cable to the ship's superstructure, is clean and tight.

An indication of possible transmit antenna or feeder problems may be gained from the reading on the voltage standing wave ratio (VSWR) meter, if one is fitted. This simple device, fitted in series with the feeder cable, measures the ratio between forward power to the antenna and the reflected power back down the feeder from the antenna. The reading, as a ratio of the two, would in theory be 1 (or unity). However, figures between 1.2 and 1.5 are good with an acceptable figure of 2 considered to be maximum. A figure of 6, for instance, indicates that approximately 50% of the power applied to the transmitting antenna is being reflected back down the feeder to the transmitter. A considerable waste of power indeed.

3
GMDSS personnel qualifications

3.1 Introduction

Every ship operating within the GMDSS must carry personnel qualified to operate the GMDSS equipment to provide distress and safety radiocommunications.

There must be at least one person on-board holding a GMDSS operators certificate of competency. He/she will be responsible for radiocommunications during distress incidents.

There are two types of operator certificate as defined by the IMO.

The Restricted GMDSS Operators Certificate

The GMDSS Restricted Certificate states that the named holder has passed an examination in the following subjects:

- practical knowledge of the adjustment of radio-telephone equipment;
- practical knowledge of radio-telephone operation and procedures;
- the sending and receiving of spoken messages by radio-telephone; and
- a general knowledge of radio-telephony regulations, particularly those regulations relating to safety of life at sea.

The examination consists of both oral and practical tests as follows. The candidate:

- operates radio-telephone equipment, including varying the transmitter power and changing the transmission frequency;
- possesses a basic knowledge of radio-telephone procedures with particular emphasis on distress procedures;
- maintains the batteries;
- maintains the radio communications log; and sends and receives messages correctly by radio telephone.

The General GMDSS Operators Certificate

The GMDSS General Operators Certificate (GOC) states that the named holder has been examined, by written, oral and practical means, and passed in the following subjects.

A. Knowledge of the Maritime Mobile Service and the Maritime Mobile Satellite Service.

    A.1. The general principles and the basic features of the MMS.

    A.2. The general principles and the basic features of the MMSS.

B. Detailed practical knowledge and the ability to use the basic equipment of a ship station.

B.1.   Use in practice the equipment of a ship station.

B.2.   DSC.

B.3.   General principles of NBDP and Telex Over Radio (TOR) systems. Use maritime NBDP and TOR equipment in practice.

B.3.   Usage of Inmarsat systems. Use Inmarsat SES or simulator in practice.

B.4.   Fault locating.

C.   Operational procedures and detailed practical operation of GMDSS systems and subsystems.

C.1.   GMDSS.

C.2.   INMARSAT.

C.3.   NAVTEX.

C.4.   EPIRBs.

C.5.   SARTs.

C.6.   Distress, urgency and safety communication procedures in the GMDSS.

C.7.   SAR.

D.   Miscellaneous skills and operational procedures for general communications.

D.1.   Ability to use the English language, both written and spoken, for the satisfactory exchange of communication relevant to the safety of life at sea.

D.2.   Obligatory procedures and practices.

D.3.   Practical and theoretical knowledge of general communication procedures.

A detailed examination syllabus produced by the CEPT for the GMDSS GOC examination can be found in Appendix 1.

The designated radiocommunications personnel on ships trading in A1 areas only must hold at least the Restricted Certificate. For vessels in all other areas at least one person must hold the GMDSS General Operator Certificate.

Maintenance qualifications

Two standards of certificates for the maintenance of radio equipment at sea exist within the GMDSS. They are:

(a) At-sea Maintenance Certificate Class 1 (ASM1)
(b) At-sea Maintenance Certificate Class 2 (ASM2).

The knowledge required to obtain either of these certificates involves an understanding of the technological descriptions of the systems and the maintenance techniques necessary to keep the on-board GMDSS equipment operating in peak condition.

The syllabus content of the ASM1 certification contains communications principles and systems knowledge to a greater depth than ASM2.

A number of national authorities have modified existing qualifications to satisfy GMDSS regulations regarding on-board maintenance. Details of syllabuses and examinations for these qualifications may be obtained from maritime colleges throughout the world.

Section Two
GMDSS SATELLITE COMMUNICATIONS

Introduction

This section of the book is of necessity the largest of the three. Since the inauguration of space programmes by a number of the world's leading technological countries, there has been a consistent expansion of satellite communication technology in order to satisfy the increasing demands of global communications. In order to continue to satisfy the ever expanding needs of global communications, the technology employed must constantly strive to be one step ahead of demand. This unchanging situation has lead to a massive growth of those industries which support satellite communications systems with a consequent expansion of the technology employed. The GMDSS has benefited greatly by using the latest technology, developed and financed in many cases, for military applications.

Chapter 4 contains outline satellite communication system principles. This chapter provides the reader with the necessary basic information to understand the reasons why satellite communication methods and systems have been employed in the GMDSS radionet. In addition, many of the standard terms used in satellite communications and their significance are explained. Readers wishing to research the subject of satellite communications in greater detail should read Chapter 4 and refer to the companion volume *Satellite Communications: Principles and Applications*.

4
Satellite orbital parameters and outline satellite communication principles

4.1 Introduction

Whilst the subject of satellite orbital parameters is extremely complex and may appear to the reader to be somewhat academic when considering the operation of GMDSS equipment, it will aid system understanding if the rudimentary principles of satellite orbits are considered. The space segment of the GMDSS currently includes two types of satellite each with a very different type of orbit. GMDSS distress alerting may be achieved using the COSPAS/SARSAT polar orbiting satellites which are only accessible for alerting purposes as the satellite passes within range of an EPIRB, and the geo-stationary orbiting (GEO) satellites which, as part of the Inmarsat organization, provide constant earth coverage for radio communications.

Polar orbit

As the term suggests, a satellite in polar orbit will travel its course over the geographical North and South Poles and will effectively follow a line of longitude. However, it must be remembered that the earth is revolving below the orbit and consequently the satellite will pass over any given point on the earth's surface.

The orbit may be virtually circular or elliptical depending upon requirements. If two such satellite orbits are spaced at 90° to each other the time between satellite passes over any given point will be halved. More satellites in orbit will reduce the time even further. The parameters concerning this type of orbit are fully described in the chapter entitled 'Navigation by Satellite' in the companion volume *Electronic Aids to Navigation (Position Fixing)*. A polar orbiting satellite is rarely used for communication purposes because it is in view of a specific point on the earth's surface for only a short period of time. Also, complex steerable antenna systems would be needed to follow the satellite as it passed overhead. Such a satellite is used in conjunction with omnidirectional antennae for navigation purposes and for distress alerting in the GMDSS and has been fully described in Chapter 1.

Elliptical inclined orbit

The inclination of a satellite orbit is the angle which exists between that orbit and the earth's equator. Thus, the geostationary orbit is 0° inclined whilst the polar orbit is 90° inclined. In practice, a satellite can follow an orbit with any angle of inclination. Actually, many satellite orbits are inclined by accident because it is not easy to launch satellites into pre-determined orbits.

In future, Inmarsat will use a combination of satellites in geostationary and elliptical orbits to provide communications for mobile earth stations.

Fig. 4.1   A USAF DELTA rocket carries an Inmarsat-2 F2 satellite into orbit. (Courtesy *Ocean Voice*)

Circular equatorial orbit (GEO)

A satellite placed into circular equatorial orbit may be made to appear stationary when viewed from the earth's surface. Such an orbit may be called a geostationary orbit, and satellites in that unique circular orbit become geosynchronous satellites. They are satellites whose orbital period is synchronized to the period of rotation of the earth's surface relative to their distance from it. By the use of three satellites equally spaced in geostationary orbit it is possible to provide continuous communications from one point on the earth to another. Satellites in this orbit handle the bulk of military and

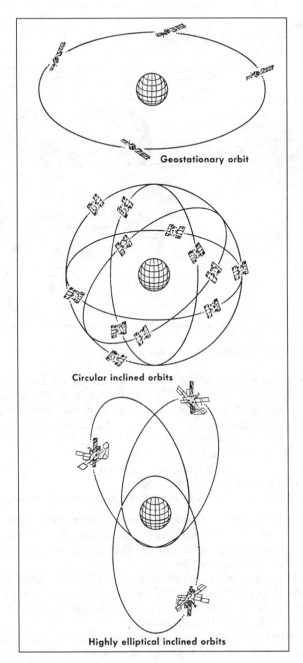

Geostationary orbit

Circular inclined orbits

Highly elliptical inclined orbits

Fig. 4.2   Satellite constellations for aeronautical mobile services. (Courtesy *Ocean Voice*)

commercial international communications on the earth. In addition, geostationary satellites are of prime importance within the GMDSS radionet as they are able to provide reliable communications from a mobile earth station to shore via the Inmarsat system.

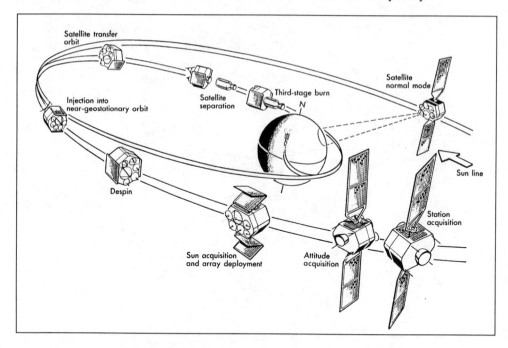

Fig. 4.3   Typical launch to geostationary orbit by expendable vehicle. (Courtesy *Ocean Voice*)

In October 1945 the science fiction writer and noted engineer Arthur C. Clark, published an article in the magazine *Wireless World* in which he calculated that a geosynchronous orbit would exist directly above the earth's equator at a distance of some 35,855 km. To maintain a position in this orbit a satellite must travel at a velocity of 3073 km s^{-1} and possess an orbital period of 23 hours, 56 minutes, 4 seconds—called one sidereal day. Clarke's 1945 article effectively formed the basis of the international communications network which is enjoyed today and, in his honour, the geosynchronous orbit is often called the Clarke Orbit.

The advantages of using satellites in the geostationary orbit are numerous.

- Fixed ground station antennae need not be steerable and consequently could be made to a simpler design. Omnidirectional antennae are now being produced, as used in the Inmarsat-C ship earth station, thus eliminating the need for complex gyroscopic satellite tracking systems.
- There is no relative movement between the satellite and a fixed earth station and consequently no Doppler frequency shift of the communications frequency is introduced. When using polar orbiting or elliptical satellites the Doppler frequency shift introduced can lead to reception problems.
- Continuous, instantaneous communications, between a mobile and a fixed station, are possible because the satellite is always in view.

The main disadvantage of the GEO orbit is that its altitude is very large and a considerable amount of thrust is necessary to lift a payload into orbit. In practice, an easterly rocket launch is used in order to take advantage of the earth's velocity and a point as close to the equator as possible is used as the launch site. Earth velocity is greatest at the equator. The satellite is first launched into a highly elliptical orbit whose point of apogee (the furthest point from the earth surface) is equal to the altitude required for a geostationary orbit. When the satellite reaches its point of apogee, an 'apogee motor' is fired to cause the satellite to be pushed into the new orbit.

Basically, a satellite remains in orbit when two forces, one caused by the gravitational pull of the earth and the other by the centripetal acceleration due to its angular velocity, are in balance.

The velocity (v) of a geostationary satellite should ideally be zero relative to the earth's surface, although small variations do occur. Satellite orbital velocity is 3073 km s^{-1} to enable the satellite to maintain geosynchronism. Small orbital variations occur due to the influence of other heavenly bodies but they are of no consequence to the reader as their effects are counteracted in the ground control station.

The altitude of the geostationary orbit may be readily calculated.

For a circular orbit at altitude (h) above the earth's equator, the circumferential path is given as:

$$2\pi(a + h)$$

where a = the average earth radius (6371 km)
$\quad\quad h$ = altitude above the earth's surface.

The circumferential velocity (v) is constant, therefore the period of one orbit is:

$$T = \frac{2\pi(a + h)}{v}$$

All satellites maintain their orbits with reference to velocity, mass and earth gravity. The centripetal force on a satellite with a mass (m) is:

$$\frac{mv^2}{(a + h)}$$

and the earth's gravitational pull is the product of mass and gravity (mg), where the gravitational force $g = 9.81$ m/s. The gravitational acceleration g' is therefore:

$$g' = g\left(\frac{a}{a + h}\right)^2$$

Balancing centripetal force against gravitational force in order to maintain orbit:

$$mg\left(\frac{a}{a + h}\right)^2 = \frac{mv^2}{a + h}$$

$$v = a\sqrt{\frac{g}{a + h}}$$

Substituting v above into the orbital period T formula and including numerical values gives:

$$h = (5075T - 6371) \text{ km}$$
$$h = 35{,}855 \text{ km (for a 24 hour period)}.$$

Angle of elevation

This is the angle between a tangent drawn to the visual horizon from a given point on the earth's surface and the direct line-of-sight path to a satellite from that same point.

If the observer is standing on the equator directly beneath a geostationary satellite, the angle of elevation will be a maximum of 90°. If the observer now moves in any

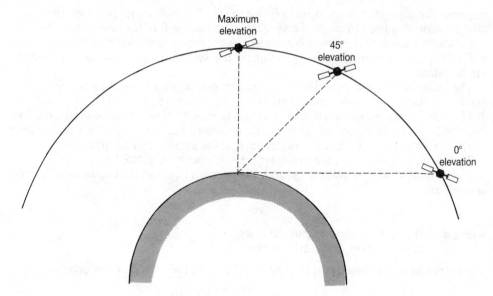

Fig. 4.4 Indications of the satellite angle-of-elevation with respect to the surface of the Earth and the corresponding range change

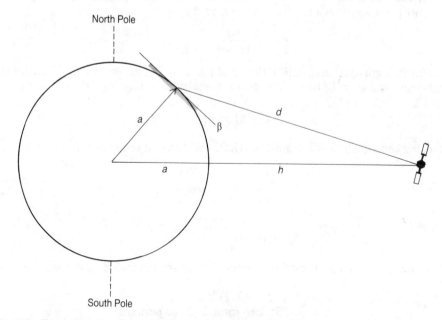

Fig. 4.5 Transmission path range with respect to the centre of the Earth

direction away from that point to the outer edge of the satellite footprint coverage area, the angle of elevation decreases until it will ultimately be zero. However, in practice the absolute minimum angle of elevation for a commercial signal path is approximately 5°. The amplitude of the received signal will progressively decrease as the observer moves away from the 90° elevation point.

Transmission path range variation and path loss

$$\text{The } \angle \text{ of elevation} = \beta$$
$$\text{By applying cosine rule;}$$
$$(a + h)^2 = a^2 + d^2 - 2ad(\cos 90 + \beta)$$
$$= a^2 + d^2 - 2ad \sin\beta$$
$$d = \sqrt{(a + h)^2 - (a \cos\beta)^2} - a \sin\beta$$
$$\text{When the observer is beneath the satellite } \beta = 90° \ (\cos\beta = 0, \sin\beta = 1)$$
$$d = \sqrt{(a + h)^2} - a = a + h - a$$
$$\therefore d = h = 35{,}855 \text{ km}$$
$$\text{When the observer is in a position where } \beta = 0°$$
$$d = \sqrt{(a + h)^2 - a^2} - 0 = 41{,}745 \text{ km}$$

Because signal attenuation increases with distance it is obvious that the received signal will become progressively weaker as the angle of elevation decreases.

Almost all of the transmission path loss of signal amplitude occurs because of spreading of the transponder beam over a large area. For a given amplitude of signal at the satellite transponder, the larger the footprint created on the earth's surface the greater will be the transmission path loss. The transmission path loss may be approximated as:

$$L = 32.5 + 20 \log d + 20 \log f \qquad \text{dB}$$

where, d = distance in km between transponder and receiver, and,
   f = transmission frequency in MHz.

All signals from a satellite must obviously travel through the earth's ionosphere and troposphere, both of which will absorb and scatter signal energy, leading to signal loss. Transmission signal path losses caused by each of these natural phenomena are proportional to the path length within the medium. The path length will be much greater at low elevations, rising to a maximum at 0° elevation leading to the maximum signal loss due to this factor. In practice, when the angle of elevation distended by a receiver is 90° the radiowave will travel directly through the stratosphere and ionosphere with a path length through the medium of approximately 120 km, whereas when the angle of elevation is 0° the path length increases to approximately 720 km.

This sixfold increase in path length within the attenuation medium leads to a corresponding decrease in signal strength.

Signal attenuation due to atmospheric effects varies inversely with frequency. In addition, the atmosphere possesses two natural absorption peaks, one at 60 GHz due to the oxygen molecule effect, and the other at 22.2 GHz caused by excessive water vapour. Both of these frequencies are above the bands currently used for commercial satellite communications and are of little consequence to the reader.

Satellite footprints

As figure 4.6 illustrates, the footprint describes the area of usable signal strength for a given beam width. The beam of a satellite downlink transmission may be wide and produce an almost circular footprint of effectively 1/3 earth coverage. Alternatively, the transponder beam may be designed to be narrow and shaped to cover a specific region, as used by television broadcasting and some communication services. The gain figures, shown in figure 4.6, provide an indication of the drop in signal strength evident as the receiving antenna is positioned away from the centre of the beam.

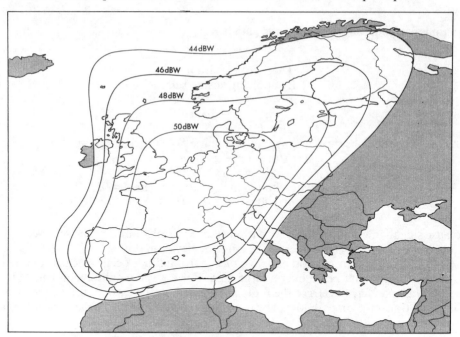

Fig. 4.6 Gain contours for one of the transmit beams of the Eutelsat 2 regional fixed sat-coms spacecraft. The beam is shaped to concentrate maximum power on western Europe. (Courtesy Eutelsat/*Ocean Voice*)

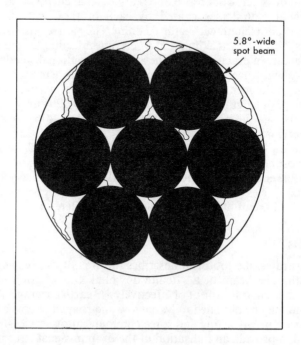

Fig. 4.7 Earth coverage by seven spot beams. (Courtesy *Ocean Voice*)

In the future, narrow elliptical beams from geostationary satellites will be used to produce footprints as shown in figure 4.7. The main advantage with this system is that specific earth areas can be covered more accurately than with a wide beam. Also, greater power can be concentrated in a smaller area, when compared with that produced by a circular beam, leading to the possible use of smaller receiving antennae.

4.2 Satellite link parameters

Introduction

Consider a satellite in geostationary orbit linked to earth stations, as shown in figure 4.8. The satellite receives an information signal from the transmitting station at a specific carrier frequency on what is termed the up-link. The satellite then amplifies this received signal and translates its carrier frequency before transmitting it to the receiving earth station via the down-link. Such a system is called the space segment, while the earth station, together with its terrestrial links, is called the ground segment.

Obviously, a satellite requires power for its operation. This power is derived from solar cells placed strategically around the satellite which convert solar energy to electrical energy. Batteries allow power to be maintained when a satellite is in earth shadow, during which time solar energy cannot be absorbed. A stabilization system is fitted which keeps the satellite fixed in its required orbital position and inclination with respect to the earth. Corrections for longitudinal and inclinational drift can be made using thrusters. Although these corrections are required only infrequently, the thrusters use fuel to obtain the corrections and the amount of fuel that can be carried is one of the factors to be taken into account when satellite operational lifetime is predicted. Transponders receive the up-link signals, translate the frequency and, after amplification, retransmit the signal to the receiving earth station. Additionally, there is a Telemetry, Tracking and Command (TT&C) system which operates in parallel with the communi-

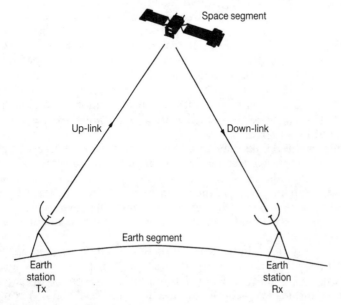

Fig. 4.8 Satellite communications links

cations system via the transponder. TT&C is for the reception of commands from the satellite control earth station and for relaying information about the satellite to the earth station. Reception of a signal and its retransmission could be via a single antenna (although there are usually many antennae fitted to a satellite). A single antenna may have a global beam coverage or, by making the antenna more directive, the beam could give a smaller zone beam coverage, or still smaller spot beam coverage. The transponders, antennae and all associated electronics comprise what is called the communications payload.

The earth station may be required to have transmit *and* receive facilities, as in the case of telephony or telex, or have only transmit *or* receive facilities, as in the case of, say, a television broadcasting service. An earth station requires an antenna which is normally very large and requires a drive and tracking system in order that signal lock may be maintained with the satellite. Electronic system installation includes high-power amplifiers (HPAs) on the up-link and low-noise amplifiers (LNAs) on the down-link. In addition, equipment is needed for signal processing and for supervisory and control purposes. Finally, there must be an arrangement for interfacing the earth station with terrestrial communication systems and concentrating the many user links on to the satellite link.

A satellite communication system using geostationary satellites is designed to provide a cost effective service taking into account the total cost of provision and operation of the service. The technical requirements, in terms of types of service (i.e voice, data or TV), number of channels, frequencies used etc, must also be addressed by the system designer. As with all radio communication systems, frequency bands are strictly allocated for satellite communication systems. Satellites using similar frequency bands must be sufficiently separated in space to avoid possible interference problems. The choice of frequency within a band is a compromise bearing in mind the need to minimize interference and provide a suitable bandwidth for the service. As a consequence of system principles, the available system bandwidth is less when low-frequency bands are used than at higher frequencies. Additionally, there is a limit on the higher frequency band that can be used since propagation difficulties exist above about 10 GHz.

A major factor in the selection of a frequency band for communication purposes is signal attenuation. This is not so significant, however, for a satellite communication path with an elevation angle of 5° or higher at frequencies below about 10 GHz. This is because of the short path length through the earth's atmosphere and the low attenuation value per kilometre for frequencies under 10 GHz. The effect of sky noise is reduced at low frequencies so that systems with low-noise temperatures are possible at those frequencies. Many geostationary satellites use fixed station-to-satellite links in the frequency band 6/4 GHz, which is at the lower end of the allocated frequency spectrum, for the reasons outlined above.

Whereas pure terrestrial links may utilize omni-directional antenna, at least in part of the link, the use of microwave frequencies for satellite communications has the advantage of allowing the radiated signal to be concentrated into a narrow beam for transmission to the satellite. In turn, the satellite can retransmit the signal to earth in a beam shaped to provide a desired footprint for a particular requirement. Narrow and shaped beams require directional antenna, hence the familiar dish antenna used in the earth stations and satellites.

Where many earth stations share a satellite there are numerous interconnecting paths. Since some of the paths may be through a single transponder the capacity of the transponder must be shared between the earth stations. This process of sharing is known as multiple access and may be achieved in practice, in several ways. A communication system must be designed to meet a certain minimum performance specification, within the limitations of transmitter power and signal bandwidth. One important performance criterion is the signal-to-noise ratio (S/N) within a baseband frequency. Because the strength of a baseband signal received by the satellite, or the earth station, is small and comparable in size with noise levels, it is essential that the signal can be recovered in the presence of the noise and this will determine the minimum signal-to-noise ratio in a receiver baseband channel. The carrier-to-noise ratio (C/N) of the RF or IF signal in the receiver will affect the signal-to-noise ratio in a baseband channel. Other factors which must be taken into account include the type of modulation used to impress the baseband signal on to the carrier and the IF band baseband channel bandwidth in the receiver. For the design of a satellite communication link it is necessary to calculate the carrier-to-noise ratio; this is illustrated in the next section.

Because of the huge propagation distances involved, the use of telephony via a satellite link will cause a delay to be produced in the link. Radio signals travel at the speed of light and, allowing for the time needed for processing the signal on board the satellite, a time delay of greater than 300 ms is experienced via a geostationary satellite link. For a two-way link a user would have to wait in excess of 600 ms to hear a response. This is much greater than the delay time experienced on terrestrial links, which typically averages out to around 30 ms. This delay may affect voice circuits if echo is also present. Echo control devices are available which assist in minimizing the echo effect. The time delay also affects digital communications and the use of forward error correction (FEC) may be necessary to overcome the effects.

Power considerations

An isotropic transmitter, in free space, radiates power uniformly in all directions. At a receiving station the power is expressed as a power flux density (PFD) which is defined as the ratio, in decibels, of RF power with respect to $1 \, \mathrm{W \, m^{-2}}$ in a specified RF bandwidth. In essence, the greater the flux density at the receiving antenna the better the signal strength and quality will be prior to processing in the receiver. The power flux density may be measured at any distance from the source in terms of the radiated power and the distance from the source. Power flux density bears an inverse ratio to distance and, consequently, will decrease as the distance increases between transmitter and receiver.

Practical satellite systems use directive antennae which cause the radiated power to exist in a given direction. The gain of the antenna in a specified direction is defined as the ratio of power per unit solid angle radiated in a given direction, to the average power radiated in a unit solid angle. In other words, concentrating the power in a given direction can increase the power flux density available at a given distance compared with an isotropic radiator of the same source power. Directivity of the antenna is of course essential in order to complete the link between earth station and satellite successfully when both systems use directive antennae. The added bonus of increased power flux density allows lower values of source power to be used than would otherwise be the case; this is especially important in satellite design when the payload is being considered and battery power is finite.

There are losses associated with the propagation of power in space such as attenuation, which has been mentioned previously.

Other losses which need to be considered are absorption and scattering of the signal. In addition, the antenna has an aperture efficiency which accounts for all losses between the incident wavefront and the antenna output port. Losses include the illumination efficiency of the antenna, phase errors, mismatch losses etc. The value of aperture efficiency varies, being in the range 50–70% for large parabolic reflector antennae; lower for smaller antennae and higher for large Cassegrain antennae. For horn antennae the aperture efficiency can have values of about 90%.

The power received by an antenna can be calculated in terms of the source power, the gain of the transmitting antenna, power losses in the transmission medium, distance of the link, gain of the receiving antenna and antenna efficiency. One of the terms frequently encountered in descriptions of a satellite system is the effective isotropic radiated power (EIRP) which is simply the product of the power transmitted and the gain of the transmitting antenna. A system designer is thus able to vary link parameters in order to achieve a target value of received power.

Most satellite links use a modulation method, for analogue and digital systems, in which the carrier amplitude remains unchanged by the modulation process. Because of this, for such links the carrier power (C) is the same value as received power.

System noise

In a conducting medium there will be a random movement of charge carriers which, assuming other factors remain constant, will produce a noise power. The available noise power delivered into a matched resistive load (the antenna) is independent of frequency and is known as 'white' noise.

It can be shown that the power delivered to a matched resistive load is independent of the value of resistance. Because the random movement of charge carriers in a resistance depends on the temperature at which the system operates, the noise power delivered by any source in a system can be defined as thermal noise measured in terms of noise temperature. Noise temperature is a useful method of determining the amount of thermal noise generated by devices in a system. At microwave frequencies all elements with a physical temperature greater than zero degrees Kelvin will generate noise at the system frequency within the system bandwidth.

The communications signals carried over a satellite system inherently have to traverse large distances, thus suffering considerable attenuation resulting in very low signal strength at the receive end of the link. Reduction of noise level is important in order to achieve good carrier-to-noise ratios and one contribution towards this is by designing the receiver bandwidth, usually in the IF stages, such that only the signal and immediate sidebands are accommodated.

Because of its effect on the noise levels in a receiver, systems are designed to keep the noise temperature as low as possible. The use of GaAs FET amplifiers, or uncooled parametric amplifiers, allows lower noise temperatures to be achieved. Even better values are obtainable if the front-end amplifier is cooled to keep its physical temperature low; this is possible in large earth stations but is costly and not feasible for mobile earth stations.

In order to measure the overall performance of a receiver it is necessary to know the total thermal noise against which the signal must be demodulated. For a detailed description of this the reader should refer to the companion volume *Satellite Communications: Principles and Applications*.

G/T ratio for earth stations

A critical parameter often quoted in the technical specifications of satellite communication links is known as the G/T factor. The expression G/T (where G is the gain of the receiving antenna and T is defined as that noise temperature which, applied to the input of a noiseless receiver, produces the same noise power output as the actual noisy receiver) is a factor, sometimes called a figure of merit, widely quoted for satellite receiving systems. The unit used for expressing the G/T of a system is decibels per degree Kelvin (dB K^{-1}), but since T is often quoted in dB relative to 1 K, or dBK, the ratio of G/T may be referred to as dBK. The Inmarsat-B specification gives minimum receive system G/T for a CES at C-band (4/6 GHz) as 32 dBK or 30.7 dBK depending on the operational satellite. The full specification would include the elevation angle (5° for the Inmarsat-B system) since temperature (T) depends on sky-noise temperature, which increases as the elevation angle is reduced below 10°.

G/T values may be low or even negative in value. The Inmarsat-B specification, for example, quotes a minimum receive antenna system G/T of $+2$ dB for CES with an L-band (1.5/1.6 GHz) capability where reception of inter-station signalling channels and the C-to-L band AFC pilot is required. The Inmarsat-B system SES has a value of G/T equal to or greater than -4 dBK in the direction of the satellite. A negative value of G/T simply means that the numerical value of T is greater than the numerical value of G.

4.3   Link budgets

Introduction

A satellite system must be planned in order to ensure that the transmission link can be satisfactorily established and maintained, having regard to the conditions under which the link is operated. A link will experience various gains and losses and collecting the gains and losses together produces what is known as the link budget.

Such a budget may be defined in terms of an up-link from, say, a CES or SES to a satellite and a down-link from the satellite to the CES or SES. Alternatively, the link budget may be defined in terms of a forward link, which may be from a LES, say, to a MES, and a return link which sends the signal from the MES, via the satellite, to the LES.

As an example, details of the Inmarsat-B system are given for voice (V), data (D) and the Network Coordination Station (NCS), time division multiplex (TDM) channels. A typical arrangement for the system is shown in figure 4.9.

Frequencies used for the CES-satellite links are in C-band with 6 GHz for the up-link and 4 GHz for the down-link. Frequencies used for the SES-satellite are in L-band with 1.6 GHz for the up-link and 1.5 GHz for the down-link.

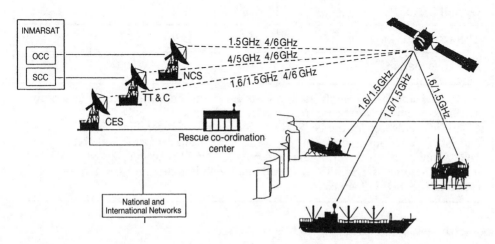

Fig. 4.9   The Inmarsat-B satellite communications infrastructure

INMARSAT-B SHORE-TO-SHIP LINK BUDGET PARAMETERS				
Channel Type.	Unit.	CESV	CESD	NCS/CES TDM *
CES elevation angle.		5 deg.	5 deg.	5 deg.
CES EIRP.	dBW	56.0	56.0	51.0
Path Loss.	dB	200.9	200.9	200.9
Absorption loss.	dB	0.4	0.4	0.4
Satellite G/T.	dBK	− 14.0	− 14.0	− 14.0
Mean up-path C/No.	dBHz	69.3	69.3	64.3
Mean satellite C/IMo.	dBHz	60.7	60.7	55.7
SES elevation angle.		5 deg.	5 deg.	5 deg.
Satellite EIRP.	dBW	16.0	16.0	11.0
Path loss.	dB	188.5	188.5	188.5
Absorption loss.	dB	0.4	0.4	0.4
SES G/T.	dBK	− 4.0	− 4.0	− 4.0
Down-path C/No.	dBHz	51.7	51.7	46.7
Nominal unfaded C/No.	dBHz	51.2	51.2	46.2
Fading margin.	dB	4.0	4.0	4.0
Overall C/No.	dBHz	47.2	47.2	42.2
Theoretical C/No.	dBHz	45.9	45.2	39.4
Available margin.	dB	1.3	2.0	2.8

Note: The above link budgets apply to Inmarsat first and second generation satellites assuming worst-case transponder conditions.
* This link budget is for the NCS TDM channel used for combined NCSA, NCSC and NCSI functions and also when used for separate (or combined) NCSA and NCSC functions. However, if it is used only for NCSI function the "SES G/T" of − 4.0 dB become "CES G/T" of 2.0 dB and the fading margin 0 dB. Similar is the case for CES TDM channels.

Fig. 4.10   Inmarsat-B shore-to-ship link budget parameters

INMARSAT-B SHIP-TO-SHORE LINK BUDGET PARAMETERS				
Channel Type.	Unit.	SESV	SESD	SES TDMA & SESRQ/SESRP
SES elevation angle.		5 deg.	5 deg.	5 deg.
SES EIRP.	dBW	33.0	33.0	33.0
Path loss.	dB	189.0	189.0	189.0
Absorption loss.	dB	0.4	0.4	0.4
Satellite G/T.	dBK	−11.2	−11.2	−11.2
Mean up-path C/No.	dBHz	61.0	61.0	61.0
Mean satellite C/IMo.	dBHz	63.7	63.7	63.7
CES elevation angle.		5 deg.	5 deg.	5 deg.
Satellite EIRP.	dBW	−5.5	−5.5	−5.5
Path loss.	dB	197.2	197.2	197.2
Absorption loss.	dB	0.4	0.4	0.4
CES G/T.	dBK	32.0	32.0	32.0
Down-path C/No.	dBHz	57.5	57.5	57.5
Nominal unfaded C/No.	dBHz	55.2	55.2	55.2
Fading margin.	dB	4.0	4.0	4.0
Miscellaneous loss.*	dB	1.5	1.5	1.5
Overall C/No.	dBHz	49.7	49.7	49.7
Theoretical C/No.	dBHz	45.9	45.2	45.2
Available margin.	dB	3.8	4.5	4.5

* Note: including loss due to SES HPA impairment and adjacent channel interference. The above link budgets represent the worst case for Inmarsat first generation satellites. Second generation satellite link budgets have the following differences resulting in improved overall link performance.
– satellite G/T = − 12.5 dBK at edge of coverage.
– satellite C/IMo = 65.1 dBHz.
– satellite EIRP = − 3.1 dBW (resulting from improved transponder gain).
– path loss = 195.9 dB on C-band down-link.
– CES G/T = 30.7 dBK

Fig. 4.11   Inmarsat-B ship-to-shore link budget parameters

INMARSAT-B CHANNEL PARAMETERS						
Type.	FEC rate.	Channel rate. kbts/s.	SES elev-ation angle.	BER (99% time).	E_b/No * (dB).	C/No * (dBHz)
CESV/SESV Telephony (including voice-band data up to 2400 bits/s)	3/4	24	10° 5°	10^{-4} 10^{-2}	4.7 3.3	47.3 45.9
CEST/CESDL Forward telex and low-speed data.	1/2	6	5°	10^{-5}	4.6	39.4
SEST/SESDL Return telex and low-speed data.	1/2	24	5°	10^{-5}	4.4	45.2
CESD/SESD 9.6 kbits/s SCPC data.	1/2	24	5°	10^{-5}	4.4	45.2
CESA/NCSC/ NCSA/NCSS Forward signalling.	1/2	6	6°	10^{-5}	4.6	39.4
NCSI/CESI Interstation links.	1/2	6	5°	10^{-5}	4.6	39.4
SESRQ/SESRP Return signalling.	1/2	24	5°	10^{-5}	4.4	45.2

* Note: Theoretical value required in order to achieve the BER value under Additive White Gaussian Noise conditions.

Fig. 4.12   Inmarsat-B channel parameters

Figure 4.10 shows a shore-to-ship link budget for CESV channels (voice telephony), CESD channels (9.6 kbits s^{-1} SCPC data/facsimile) and NCS/CES TDM channels (telex, low-speed data and signalling).

Figure 4.11 shows a ship-to-shore link budget for SESV channels (voice telephony), SESD channels (9.6 kbits s^{-1} SCPC data/facsimile) and SES TDMA/SESRQ/SESRP channels (telex, low-speed data and signalling).

The link budgets shown represent channel performance under generally worst-case satellite transponder conditions for Inmarsat first and second generation satellites (MARECS, INTELSAT-V MCS and Inmarsat–2) and are shown only as examples.

For a receiver, the power delivered to the receiver demodulator input (C) represents the power in the carrier and sidebands and the carrier-to-noise power ratio at this point is given by C/N, where N is the noise power at the demodulator input. The value of noise power at this point can be shown to be equal to kTB, where k is a constant, T is the noise temperature and B is the bandwidth of the IF filter. Prior to the IF filter of the receiver the ratio of carrier power to noise power is expressed as C/N_0, where N_0 is given by kT only. Thus, C/N is equivalent to C/N_0B; if C/N is measured in decibels (dB), the units for C/N_0 are therefore in dBHz. The values shown for 'theoretical C/N_0' for each channel type relate to the corresponding values given in Figure 4.12.

The values shown for 'available margin' in Figures 4.10 and 4.11 relate to the margin above the theoretical C/N_0 requirements after taking into account fading and other link losses. The margin therefore includes equipment implementation degradations compared with the assumed theoretical performance.

Inmarsat-B link budget

To examine a link budget in greater detail, consider the example of the Inmarsat-B link budget for a telephony channel for a satellite–ship down-link and a ship–satellite up-link, using the INTELSAT-V MCS satellite.

(a) *Satellite–ship down-link*

Frequency	1535–1542.5 MHz
Satellite EIRP	16 dBW
Path loss at 1542.5 MHz at 40 000 km range	
$L_p = 20\log[4\pi d/\lambda]$	188.5 dB
Absorption loss (L_a)	0.4 dB
SES G/T	-4 dBK
Boltzmann's constant (k)	-228.6 dBW^{-1} Hz^{-1} K^{-1}
$C/N_0 = (\text{EIRP} - L_p - L_a + G/T - 10\log k)$	51.7 dBHz

Figure 4.10 gives the down path C/N_0 as 51.7 dBHz. A nominal value of 51.25 dBHz together with a fading level of 4 dB gives an overall value of C/N_0 of 47.2 dBHz. By comparison with the theoretical value of 45.9 dBHz required to give the BER value under white Gaussian noise there is a margin of 1.3 dBHz.

(b) *Ship–satellite up-link*

Frequency	1636.5–1644 MHz
Ship terminal EIRP	33 dBW
Satellite G/T_s	-11.2 dBK
Path loss to satellite at 1636.5 MHz	
$L_p = 20\log[4\pi d/\lambda]$	189 dB
Absorption loss L_a	0.4 dB
Boltzmann's constant (k)	-228.6 dBW^{-1} Hz^{-1} K^{-1}
$C/N_0 = (\text{EIRP} - L_p - L_a + G/T_s - 10\log k)$	61 dBHz

Figure 4.11 gives the up path C/N_0 as 61 dBHz. The effect of intermodulation in the transponder band is given as C/IM_0 with a value of 63.7 dBHz.

A relationship exists between the maximum EIRP value and the power flux density (PFD) at the edge of earth coverage as a function of occupied transponder bandwidth. In certain types of systems only a small part of the transponder bandwidth may be occupied so as to enable the available EIRP to be concentrated into the narrower bandwidth, thus raising the EIRP density (dBW Hz^{-1}) and therefore raising the *PFD* (dBW m^{-2} Hz^{-1}). If the *PFD* value is specified, the available EIRP will set the upper limit to the amount of bandwidth which can actually be used.

The CCIR has set a limit to the maximum value of permissible flux density at the surface of the earth to prevent interference with terrestrial links due to a satellite transmitting at 4 GHz. For shared frequency channels in the band 3.4 to 7.75 GHz, the maximum permitted flux density, *PFD*$_{max}$, in any 4 kHz bandwidth slot, for a wave arriving at an angle, $\theta°$ where θ is between 5° and 35°, above the horizon, is given by:

$$PFD_{max} = [-152 + \left(\frac{\theta - 5}{2}\right)] \quad \text{dBW m}^{-2}$$

where $\theta = 0°$, *PFD*$_{max}$ = -152 dBW m^{-2}

At 4 GHz, 0° elevation, the flux density is limited to -152 dBW m^{-2} for a 4 kHz band, so if the value of *PFD* is greater than this level, energy dispersal may be required to prevent all the transponder power being radiated at one frequency if no modulation was applied to the carrier.

Assuming a constant power satellite transmitter, a single narrowband (cw) carrier has the highest *PFD* because of the concentration of energy. Modulation of the carrier, or the addition of other carriers, causes the power to be spread over a wider bandwidth. *PFD* is thus decreased while the illumination level remains constant (assuming no output 'backoff'). Often, a deliberate spectrum-spreading waveform is used to produce energy dispersal, therefore reducing *PFD* to acceptable levels.

Noise at CES input is given by the expression kTB (where k is a constant, T is the noise temperature in degrees Kelvin and B is the bandwidth). Since noise is a function of temperature, it is desirable to minimize the noise temperature of the devices used. Low-noise temperatures are possible with helium-cooled parametric amplifiers where a value of $T_n = 30$ K over a 500 GHz band is possible. (T_n is the noise temperature of the device.) To this must be added antenna noise temperature (T_a) which may be, typically, 50 K for a large antenna at 4 GHz. The effects of other noise sources within the receiver could lead to a system temperature (T_s) of, say, 180 K. Assuming a channel bandwidth of 20 kHz, the system noise can be calculated:

$$\begin{aligned} N = P_n &= kT_sB \\ &= 1.38 \times 10^{-23} \times 180 \times 20 \times 10^3 \text{ W} \\ &= 4.968 \times 10^{-17} \\ &= 10\log(4.968 \times 10^{-17})\text{dBW} \\ &= -163 \text{ dBW} \end{aligned}$$

To maintain satisfactory communications, the value of C/N must remain above a threshold value under all conditions. For a frequency modulation (FM) system, the threshold is in the range 4 to 15 dB depending on the type of demodulator used. For phase shift keying (PSK) systems, the threshold is, typically, 8 to 15 dB. Typically for the system considered, using PSK with a channel bandwidth of 20 kHz, a threshold of 10 dB is used. Additionally, there is a system margin which allows for propagation and equipment degradation. For example, at 4 GHz, if a fading of 1 dB occurs which degrades C by 1 dB, and sky noise increases to a value which increases N by, say, 1 dB, then there is a reduction in C/N of 2 dB due to propagation factors. A further allowance of, say, 1 dB could allow for equipment degradation, giving a system margin of 3 dB.

When using FM and frequency division multiple access (FDMA), the transponder cannot be operated at maximum output power because of non-linearity of the output power device—usually a travelling-wave tube amplifier (TWTA), at high values of output power. Backoff is usually employed, usually between 3 to 7 dB at the output of the device, to keep intermodulation products down to a level that is acceptable. For a system with many channels, the backoff will need to increase as the number of multiple accesses increases. Intermodulation is caused by the non-linearity of the transponder output device when operating with a multiplicity of signals. The generation of a large number of intermodulation products under these conditions can cause some interference within the transponder passband. The effect may be minor but it does cause an overall increase in noise levels which would reduce the C/N value of the required signal at the earth station.

The use of time division multiple access (TDMA) enables greater efficiency for low and high capacity stations but demands great earth station complexity. Mixing earth stations with different values of G/T is not possible using this system because each station must receive timing information with the same value of C/N.

Either frequency division multiplexing (FDM) or time division multiplexing (TDM) may be used to multiplex signals received by the earth station to provide many channels of telephone or data information. The signal received for any channel is weak and is amongst noise so that it is essential to have a good signal-to-noise ratio. A value of at least 18 dB for the S/N is required for satisfactory operation. The type of modulation used can also improve the signal-to-noise ratio for a single channel to well above the received value. The use of frequency modulation or phase shift keying could improve the value of S/N to about 53 dB in a single channel compared with the input C/N value of 18 dB.

For FM, the increase in S/N is brought about by using a high value for the frequency deviation f_d. As an example, 40 telephone channels could be sent, using FM, with an RF bandwidth of 2.4 MHz giving 60 kHz per channel. The same 40 channels could be sent, using single sideband amplitude modulation (SSAM) in an RF bandwidth of 160 kHz, using 4 kHz per channel. It follows that the improvement in S/N obtained using FM is at the expense of reduced channel capacity.

An improvement in channel capacity can be achieved with digital systems using TDM and PSK modulation. Examples which will illustrate this concept follow later in this section.

Summary

The design of a satellite communications system will endeavour to maximize the number of channels used for each transponder. Although system margins are incorporated to ensure adequate quality of received signals, an increase in capacity can be obtained if system margins and signal quality are reduced. Provided the result is not obvious to the user, the increase in channel capacity produced, by such a reduction in system margins and signal quality, can allow a significant difference to the pricing of the system. As an example of one method that has been used to increase channel capacity, consider the use of syllabic companding and time-assigned or digital speech interpolation. Since, on average, a one-way voice channel only carries speech for about 40% of the time, the channel can be allocated to another customer for part of the remaining 60% of the time. Overloads do occur in such an arrangement but since their effects on quality are brief this usually goes unnoticed by the customer.

4.4   Multiple access

Introduction

In the case of satellite communications, multiple access occurs when more than one pair of earth stations require the use of the satellite communication channel. Several transponders on the satellite share the frequency band in use and each transponder will act independently of the others to filter out its own allocated frequency band and further process that signal for retransmission. Access to a transponder may be limited to a single carrier or many carriers may exist simultaneously. The baseband information to be transmitted, whether telephony, data or video, is impressed on the carrier by the process of modulation. The modulation may be by single or multi-channel basebands.

There are many methods by which multiple access may be achieved, but only five will be briefly considered because of their relevance to the GMDSS network. The systems are as follows.

● **Frequency-division multiple access (FDMA)**

This is the separation of signals in the frequency domain. For frequency division multiplexing, each baseband signal is translated to a higher frequency by modulating a carrier frequency which then becomes the centre frequency of the signal. Other users' signals are treated in the same way so that all signals are translated to a new set of frequencies, separated from each other in order to prevent mutual interference. A group of multiplexed signals may then modulate a common carrier for transmission purposes. Such a group of multiplexed signals forms the transponder channel. Figure 4.13 shows the possible arrangement for a typical satellite system with 12 transponders, each with a channel 36 MHz wide and separated from adjacent channels by a guard band of 4 MHz. This would provide approximately 600 voice channels allowing for separation between signals within a channel, assuming analogue transmission. A means of describing multiple access systems, according to the method of assembling the signals for transmission, would give the sequence FDM/FM/FDMA for the above example.

FDM. Incoming signals are assembled into a multichannel composite signal using frequency division multiplexing techniques.

FM. The composite signal is frequency modulated on to a carrier for transmission.

FDMA. The 36 MHz bandwidth of each of the transponders is split up into segments for use by different earth stations. Each earth station having an allocated range of frequencies, within the transponder bandwidth, for its sole use.

This classification for multiple access systems will be used in the sections that follow.

The transmission could be analogue or digital in continuous or burst mode. Analogue transmission suffers badly from the effects of noise and a wider bandwidth is required for good reception. Digital transmission suffers less from the effects of noise. Using digital techniques the system described in figure 4.13 could be improved to accommodate about 800 channels.

Advantages of FDMA include: simplicity in earth station equipment; no complex timing and synchronization techniques are required, as is the case for TDMA. Disadvantages include: likelihood of inter-modulation problems with its adverse effect on the signal-to-noise ratio; and the need for fairly complex frequency plans to reduce such effects.

The example illustrated in figure 4.13 is shown for the purpose of indicating how the principle of FDMA can be applied. The figures quoted are typical for a C-band satellite system used for telephony traffic. Maritime applications to be discussed will follow the principle of FDMA but the figures for channel capacity etc will vary according to the system used.

● **Time-division multiple access (TDMA)**

Using this method a satellite transponder channel is available to a single carrier for a period of time. Each station transmits a burst of information within its specified time slot; once the burst is complete a second station transmits its information and so on. Each earth station uses the same carrier frequency but since each signal arrives at the

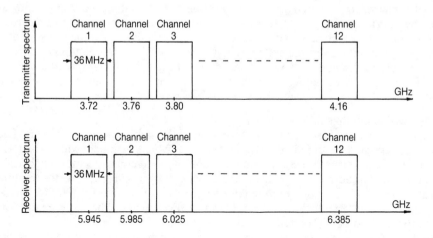

Fig. 4.13   Part of the frequency spectrum showing a basic FDMA arrangement for satellite work

satellite at different times there is no danger of interference. When signals are transmitted from the satellite to earth stations, the receiving station will select only that signal which occurs at a specified time, ignoring all others. The time slots are arranged permanently or assigned as required. Assignment of time slots can be arranged using a common control channel which can also release the station to a common pool once the transmission has been completed. TDMA is suitable only for digital transmissions. An example of TDMA is illustrated in figure 4.14.

As shown in figure 4.14 the total time slots occupied by all users constitute a frame. The frame structure repeats so that a fixed TDMA assignment is composed of time slots

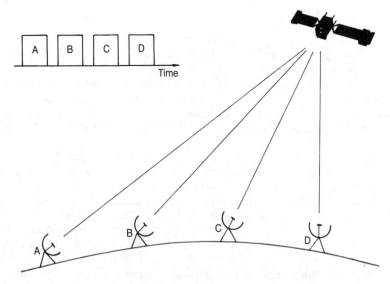

Fig. 4.14   A simple example of a TDMA system

that re-occur with successive frames. There are short time periods between successive slots, called guard times, which assist in separating successive time slot bursts and thus minimizing interference. Advantages of TDMA include: no inter-modulation problems since the transponder only has to deal with one carrier at a time, thus the transponder is able to operate at full power; the system is highly flexible, allowing channels of differing capacity to be accommodated by altering the number of equally spaced time slots allocated to a user. Disadvantages include the requirement for complex, and expensive, earth stations.

- **Code-division multiple access (CDMA)**
Users share the transponder channel and transmit at the same time as other users. Each transmission spreads its signal over a bandwidth, which is much wider than that required for the information alone, and contention is avoided because each transmission uses a unique code sequence. A receiving station can retrieve the required information by using the same unique code sequence. Advantages of CDMA include privacy and good interference tolerance. Disadvantages include the requirement for a complex earth station and poor frequency utilization.

CDMA tends to be a specialist system and is restricted mainly to military applications. It has been included here for completeness but since it is not used in the Inmarsat system it will not be discussed further

- **Packet access**
Digital signals may be organized on a packet basis. A packet may comprise traffic from several sources that has been assembled and prepared for transmission to the next point in the communications network.

- **Random access**
This is a demand assignment method with little or no central control. Access to the communications network is by contention, whereby one earth station transmits a burst of data whenever it desires access to the system, regardless of what other stations may be doing. One form of random access system is the ALOHA system developed by the University of Hawaii for Pacific Island communications.

FDMA

One possible arrangement of an FDMA system uses single carrier per channel (SCPC) whereby one signal, which could be voice or data, modulates the carrier. The modulation could be frequency modulation (FM), for analogue transmission, or phase shift keying (PSK) for digital transmission. The carrier is then transmitted using FDMA.

A second arrangement utilizes digital transmission with time-division multiplexing (TDM) to combine multiple digital channels. The digital baseband signal then modulates a digital carrier using PSK.

An advantage of using SCPC is to facilitate the use of voice-activated carriers. Such carriers are switched off during the periods between speech activity, thus reducing the power consumption. It has been established that on average a speaker will talk for only 40% of the time and switching off the carrier for the remaining 60% of the time reduces the satellite power consumption by up to 4 dB. This would enable a corresponding increase in power available and hence channel capacity for the transponder. To minimize inter-modulation between transmissions to acceptable levels, the output of the transponder power device must be backed off; typically this would be about 4 to 6 dB for a travelling wave tube amplifier (TWTA), and about 2 dB for a solid-state power amplifier (SSPA). The SSPA has a reasonably linear characteristic provided it is not over-driven. A small earth station, however, may suffer because of increased down-link noise and require a higher value of power per channel than otherwise would be the case. This would require less backoff for the transponder output power device. SCPC has the advantage that the power of individual transmitted carriers can be adjusted to optimize for particular link conditions.

SCPC requires automatic frequency control (AFC) to maintain spectrum centring on a channel by channel basis. This is usually achieved by transmitting a pilot tone in the centre of the transponder bandwidth. A receiving station uses the pilot tone to produce a local AFC system which is able to control the frequency of the individual carriers by controlling the frequency of the local oscillators

SCPC/FM/FDMA

The Inmarsat-A system uses SCPC utilizing analogue transmission with frequency modulation for telephone channels. In calculating the channel capacity of the SCPC/FM system it is necessary to ensure that the noise level does not exceed specified levels. The CCIR recommendations for an analogue channel state that the noise power at a point of zero relative level should not exceed 10,000 pWOP with a 50 dB test-tone:noise ratio. It is assumed that the minimum required carrier-to-noise ratio per channel is at least 10 dB.

SCPC/PSK/FDMA

In this arrangement, each voice or data channel is modulated on to its own radio-frequency carrier. The only multiplexing occurs in the transponder bandwidth where frequency division produces individual channels within the bandwidth.

The satellite transponder carrier frequencies may be pre-assigned or demand-assigned. For pre-assigned carriers the frequency is assigned to a channel unit and the PSK modem requires a fixed frequency local oscillator (LO) input. For demand-assignment, channels may be connected according to availability of a particular carrier frequency within the transponder RF bandwidth. For this arrangement the SCPC channel frequency required is produced by a frequency synthesizer.

A description of the various SCPC/FDMA channels of the Inmarsat-B system may be found in Chapter 7 and is illustrated in figure 7.2.

TDM/FDMA

This arrangement allows the use of a time division multiplexed group, or groups, to be assembled at the satellite in FDMA. Phase shift keying is used as the modulation process at the earth station. Systems such as this are compatible with FDM/FDMA carriers sharing the same transponder and the earth station requirements are simple and easily incorporated.

The Inmarsat-B system for telex/low-speed data uses TDM/FDMA in the shore-to-ship direction only, with the ship-to-shore direction using TDMA/FDMA. The CES TDM (and SES TDMA) carrier frequency is pre-allocated by Inmarsat. Each CES is allocated at least one forward CES TDM carrier frequency (and a return TDMA frequency). Additional allocations can be made depending on the traffic requirements.

The channel unit associated with the CES TDM channel for transmission consists of a multiplexer, differential encoder, frame (transmit) synchronizer and modulator.

At the SES, the receive part of the channel has the corresponding complementary

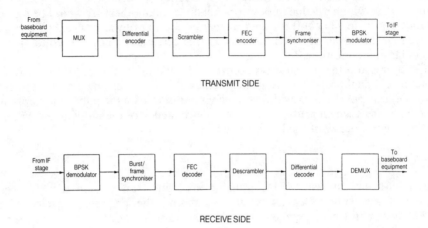

TRANSMIT SIDE

RECEIVE SIDE

Fig. 4.15 Inmarsat-B channel arrangement for TDM operation

functions to the transmit end. The full channel unit arrangement for transmit and receive channels is shown in figure 4.15.

The CES TDM channels use BPSK with differential coding, which is used for phase ambiguity resolution at the receive end.

TDMA/FDMA

As previously stated, TDMA signals could occupy the complete transponder bandwidth although this is unlikely. A variation of this is where TDMA signals are transmitted as a sub-band of the transponder bandwidth, the remainder of the transponder bandwidth being available for, say, SCPC/FDMA signals. The concept is illustrated in figure 4.16.

The use of a narrowband TDMA arrangement is well suited for a system requiring only a few channels and has all the advantages of digital transmission. Narrowband TDMA could, however, suffer from intermodulation with the adjacent FDMA channels.

Telex services of the Inmarsat-B system for shore-to-ship channels have a flexible allocation of capacity for communications and signalling slots depending on traffic

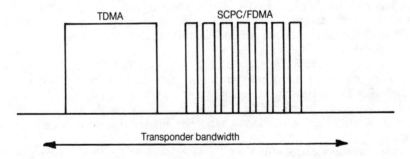

Fig. 4.16 Transponder frequency band arrangement showing part TDMA access and SCPC/FDMA access

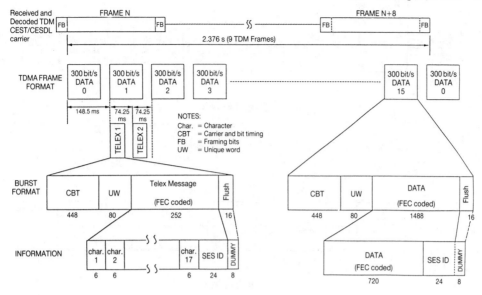

Fig. 4.17 Inmarsat-B SES TDMA frame format for telex operation

requirements. At least one slot (the number of the slot can vary) in each TDM frame is available for signalling messages. Under these conditions the maximum telex capacity in the TDM frame is 56 channels (using seven slots for telex and one slot for signalling). For ship-to-shore channels, each SES TDMA channel provides for a maximum of 32 telex bursts (SEST channel), or a maximum of 16 low-speed data bursts (SESDL channel). The SEST and SESDL functional channels cannot be combined on the same physical TDMA channel. The SES TDMA frame format for telex is shown in figure 4.17. Each SES TDMA channel is associated with a forward CES TDM (CEST) channel from which TDMA synchronization timing is derived.

The channel unit associated with the SES TDMA channel for transmission consists of a multiplexer, scrambler, FEC encoder, frame (transmit) synchronizer and modulator.

At the CES, the receive part of the channel has the corresponding complementary functions, compared with the transmit end. The SES TDMA channel uses O-QPSK modulation. The full arrangement is shown in figure 4.18.

The circuit arrangement shown in figure 4.18 is identical to that used by the Inmarsat-B system for the voice channel unit.

The Inmarsat-B telex/low-speed data channel is listed here as TDMA/FDMA for the return direction link; similarly the forward link has been identified as TDM/FDMA. This classification is reasonable since it reflects how the transponder bandwidth is allocated for the Inmarsat-B channels. However, the complete telex/low-speed data link is TDM/TDMA and is referred to as such later

TDMA

The basic concept of TDMA has been described previously. In a single carrier per transponder (SCPT) arrangement, the transponder is fully occupied by a single carrier bandwidth and the carrier is shared in time to allow several stations to transmit information, using digital modulation, in bursts. The satellite receives the earth station bursts sequentially without overlapping interference and is then able to retransmit all

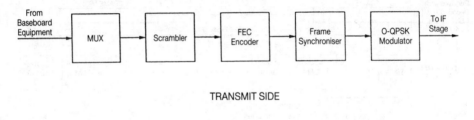

TRANSMIT SIDE

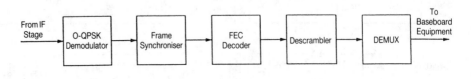

RECEIVE SIDE

Fig. 4.18   Inmarsat-B SES TDMA system showing the use of offset-QPSK modulation

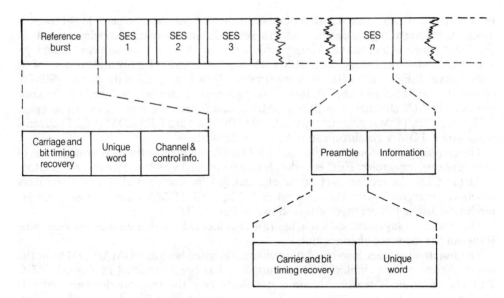

Fig. 4.19   A typical TDMA frame structure

bursts to all stations. Synchronization is necessary and is achieved using a reference station from which burst position and timing information can be used as a reference by all other stations.

So as to ensure the timing of the bursts from multiple earth stations, TDMA systems use a frame arrangement. A typical TDMA frame structure is shown in figure 4.19.

The start of a frame contains a reference burst inserted by the reference station. This is followed by additional bursts from other stations having synchronized to the reference burst to fix the timing. Each additional burst has a preamble of fixed length, carrying no traffic information, followed by the traffic information.

Because different stations possess slight differences in frequency and bit rate, the receiving station must be able to establish accurately the frequency and bit rate of each burst. This is achieved using the carrier and bit timing (CBT) recovery sequence. The form of this sequence depends on the modulation method used.

The CBT sequence is followed by a sequence known as the unique word (UW). The function of the UW is to confirm that the burst is present and to enable the determination of a timing marker that is used to establish the position of each bit in the remainder of the burst. The timing marker allows the identification of the start and finish of a message in the burst and aids correct decoding. A UW should have a high probability of correct detection.

The CBT and UW form the initial part of the preamble burst. Additional elements in the preamble could contain service information used for exchanging messages (regarding the state of the system) and control and delay channels, used for messages regarding information on acquisition, synchronization and system control.

Figure 4.19 shows the SES TDMA 24 kbit s^{-1} channel format of the Inmarsat-B system for ship-to-shore telex and low-speed data. This figure clearly shows the CBT and UW preamble sequence and indicates the number of bits involved.

TDM/TDMA

The Inmarsat-A system uses the TDM/TDMA arrangement for telex signals. Each CES has at least one TDM carrier and each of the carriers has 22, 50 baud, telex channels and a signalling channel.

In addition, there is a common TDM carrier continuously transmitted on the selected idle listening frequency by the NCS for out-of-band signalling. The SES remains tuned to the common TDM carrier to receive signalling messages when the ship is idle or engaged in a telephone call.

When an SES is involved in a telex call it is tuned to the TDM/TDMA frequency pair associated with the corresponding CES. Telex transmissions in the return direction (ship-to-shore) form a TDMA assembly at the satellite transponder. Each frame of the return TDMA telex carrier has 22 time slots; each of these slots is paired with a slot on the TDM carrier. The allocation of a pair of time slots to complete the link is received by the SES on receipt of a request for a telex call. A full description of the Inmarsat-A system can be found in Chapter 6.

There are many impairments which could affect the quality of a TDM/TDMA link, including some due to the use of TDMA bursts. These include RF signal leakage between bursts from the earth station, burst delay, improper burst bit-rate due to jitter or loss (which affects synchronization and hence clock recovery). The terminal equipment has to be able to cope with these impairments. As an example, in the Inmarsat-A system, loss of TDM frame synchronization by the SES could cause TDMA bursts to be transmitted in a time slot not assigned to the SES. A TDMA carrier transmit inhibit signal is therefore activated should there be loss of frame synchronization from the received TDM carrier

Packet access

For data transmissions a bit stream may be sent continuously over an established channel without the need to provide addresses, unique words, etc if the channel is not shared. Where sharing is implemented, data is sent in bursts which thus require unique

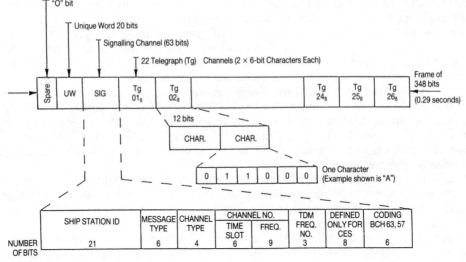

Fig. 4.20   Inmarsat-A information frame

words, synchronization signals, etc to enable time-sharing with other users to be effected. Each burst may consist of one, or more, packets comprising data from one or more sources that have been assembled, processed and made ready for transmission. An advantage of packet access, compared with the use of dedicated circuits, is that its use provides the opportunity to store data either simply to transmit at a later time or while waiting for a connection to be established. Store-and-forward methods of communication are an example of this process. Packet access can be used in random-access systems, such as ALOHA, where retransmission of blocked packets may be required.

The Inmarsat-C system incorporates automatic repeat request (ARQ) with half-rate convolutional coding with a constraint length of seven. The modulation method is BPSK. The system operates TDM/TDMA with forward channels being operated in continuous-mode TDM. Each of the channels incorporates a frame structure within which there may be a packet structure. All frames and packets are an integral number of bytes, and bits within a byte are transmitted in sequence from bit 1 to bit 8. Fields of more than a single byte are transmitted from the most significant byte to the least significant byte. The TDM channels have fixed length frames of 10368 symbols transmitted at 1200 symbols s^{-1} (600 symbols s^{-1} for first generation satellites) with a frame length of 8.64 s. Each frame has a 639 byte information field which contains consecutive packets. Any packet overlapping a frame boundary is repackaged as two 'continued' packets, one unfinished in the current frame and the remainder in the next frame. An example of an information field is shown in Chapter 8, figure 8.4. The signalling channel descriptor packets are for describing the signalling channels associated with the To-

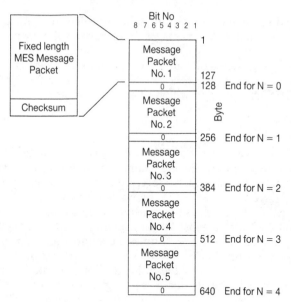

Fig. 4.21   Inmarsat-C block frame used in the packet access system

Mobile TDM. This is followed by message and signalling packets as required. If there are insufficient packets to fill the available space the remaining bytes are set to an idle value of all zeros.

The bytes of the information field are then scrambled and the scrambled data converted to a serial bit-stream. The bit-stream is then passed to a half-rate convolutional encoder which sends 10240 symbols to the interleave matrix. After assembly, the interleave block is transmitted on a row-by-row basis according to a permuted sequence. Interleaving is undertaken to improve reception of the data in the presence of channel fading.

The MES message channel is similar to the TDM channel except that, since it is quasi-continuous, a preamble is added and frame length is variable between messages. The transmission frame length is $128 + (N + 1)2048$ symbols, where N is a message block size having a value between 0 and 4. Each frame transports $(N + 1)$ message packets which are 127 byte fixed-length packets. Each packet ends with an added zero byte. Block length is thus $128(N + 1)$ bytes. The block arrangement is shown in figure 4.21.

The block is scrambled, encoded and interleaved as described for the TDM channel. Rows are transmitted in a permuted order. Empty packets, or the empty part of the final frame, are filled with zero bytes.

Random access

For random access, access to the communications link is achieved by contention. A user transmits a message irrespective of the fact that there may be other users equally in contention. Because of the random nature of the transmissions there is a possibility that transmissions from other users will collide causing the data to be blocked from receipt by the earth station. With pure random access if such a collision occurs the destination mobile station channel interface equipment would detect this and retransmit the message. The retransmissions, which could occur as many times as necessary, are carried out using random time delays. If all stations are entirely independent there is every likelihood that the original two messages that collided will be separated in time on retransmission.

Types of random access systems include ALOHA and slotted-ALOHA. Other forms of ALOHA exist such as slot-reservation ALOHA and capture ALOHA. The latter is not used in the GMDSS network and will not be discussed here.

ALOHA

This is the basic random access system which was developed in the late 1960s at the University of Hawaii to facilitate Pacific Island communications. Packets of data transmitted by random access may not be received correctly at the satellite at the first attempt and will be delayed by at least the time taken for the data to make the round trip. The originating MES also receives the transmitted data on retransmission from the satellite. This is about 0.27 s for a geostationary satellite orbiting 40,000 km above the earth. The total delay in seconds is thus 0.27 times the number of retransmissions needed before successful capture of the data. ALOHA has a low saturation capacity, typically 1/2e or 18.4%; this is an indication of the utilization of the link, where utilization is defined as the amount of time the channel is earning revenue compared with the total time. The probability of a packet being lost at the initial attempt will depend on the utilization since fewer users attempting to access the link will reduce the probability of collisions. At maximum utilization over 30% of the packets will fail to reach the satellite in recognizable form due to the high incidence of collisions.

The advantages of ALOHA are the lack of any centralized control giving simple, low-cost, stations, and the ability to transmit at any time without having to consider other users.

Slotted ALOHA

Slotted ALOHA, or S-ALOHA, is a form of ALOHA where the time domain is divided into slots equivalent to a single packet burst time. If all users transmit only at the start of a time slot then either the packet will get through or it will be in total contention with another packet. There will be no overlap as is the case with ALOHA. Figure 4.22 shows a simplified arrangement which illustrates this point.

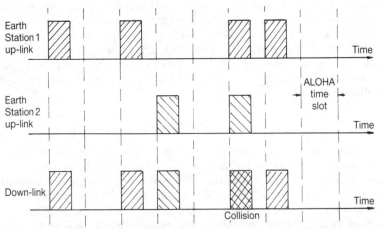

Fig. 4.22   Slotted ALOHA access showing a collision

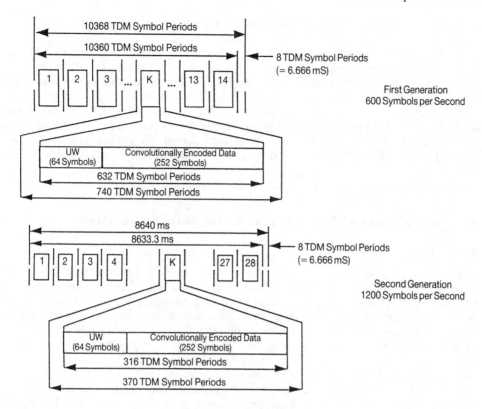

Fig. 4.23   Inmarsat-C system showing the slot access arrangement

Because of the reduced risk of collisions, S-ALOHA has a saturation capacity of 1/e, or 37%, which is twice that of ALOHA. For the same value of utilization as basic ALOHA, the time delay and probability of packet loss are both improved. The disadvantage of S-ALOHA is that more complex earth station equipment is necessary, because of the timing requirement, and that because there are fixed time slots a user with a small transmission requirement is wasting capacity by not using the time slot to its full availability. Slotted-ALOHA can only work where the transmit station has a receiver.

Slot reservation ALOHA

This is simply an extension of the S-ALOHA format whereby time slots may be reserved for a particular earth station. Slot reservation basically takes two forms.

● **Implicit.** When a station acquires a slot and successfully transmits, the slot is reserved for that station for as long as it takes the station to complete its transmission. There is a danger that a station with much data to transmit could 'lock-up' the system to the detriment of other users.
● **Explicit.** Stations may send reservation requests for a time slot prior to actually sending data. A record of all time slot occupation and reservation requests is kept. A free time slot could be allocated to a requesting station or, if all slots are occupied, the next available slot could be allocated on a priority basis.

Some control for slot reservation is necessary and this could be accomplished by a single station or by all stations being informed of slot occupancy and reservation requests.

All the Inmarsat systems use ALOHA in some form, as part of the mobile station request channel signalling requirements. As an example, Inmarsat-C uses explicit slot reservation ALOHA. The signalling channel for Inmarsat-C has the same frame length as a TDM channel (8.64 s) with each frame divided into 14 slots for first generation satellites (28 for second generation satellites). The transmission rate for a burst is 600 symbols s^{-1} (or 1200 symbols s^{-1} for second generation satellites) and the timing of the MES transmission in a slot is taken from the received To-Mobile TDM. Slots are accessed by an MES with slot bursts consisting of a unique word and data as shown in figure 4.23.

The unique word is the same as that used for the TDM (after permuting).

4.5   Modulation and demodulation for satellite services

Introduction

Baseband information to be transmitted over any radio frequency communication link consists essentially of voice, data or video signals. Each of these signals must be processed so that they are in a form suitable for transmission; this is known as baseband signal processing. The modified baseband signal is then superimposed on to a higher frequency carrier wave; the signal thus modulates the carrier and in this process undergoes frequency translation to a value suitable for propagation over the transmission link. The process of modulation at the transmit end of the link must be accompanied by demodulation, at the receive end of the link, in order to recover the baseband signal. The circuit that performs the process of modulation is the modulator while the circuit that recovers the baseband information is the demodulator. For digital systems the circuit containing a *mo*dulator for transmission, and a *dem*odulator for reception, of the RF carrier is known as a *modem*.

Where satellite links are concerned there could also be a terrestrial link involved between the user terminal and the CES or LES; this terrestrial link may be by radio link or a land line or both.

For the purposes of this section, modulation methods used for voice and data systems for transmission over a satellite RF link, using either analogue or digital transmission methods, are considered.

Telephone (voice) signals

The range of frequencies that can be received by the human ear may be approximated to about 16 kHz. However, the range of frequencies needed to produce good quality speech is less than this and is typically band limited, to a range of 300 to 3400 Hz, or even 3000 Hz, by the telephone instrument and the transmission network.

The quality of a received analogue voice signal has been specified by the CCITT to give a worst-case baseband signal-to-noise ratio for a voice signal, for transmission over a long distance, as 50 dB. Here the signal is considered to be a standard 'test-tone' and the maximum allowable noise in the baseband is 10,000 pW.

Speech is characterized by having a large dynamic range of up to 50 dB to accommodate the volume difference between a whisper and a shout. Speakers also tend to pause often while talking, giving bursts of energy of random duration and random separation. It has been found that, on average, a speaker will talk for only about 40% of the time available, the remainder of the time the link is idle.

For digital transmission, the quality of the reconstituted speech at the receive end will depend, among other factors, on the number of bits transmitted per second and the number of bits received in error (bit error rate or BER). In general, the BER necessary to give good quality speech is considered to be about 10^{-4} (1 bit error per 10,000 bits) and this value could be used as a design threshold. Some systems will have values superior to this and 10^{-5} is common.

Data signals

Data signals can be broadly classified in three ranges namely: narrowband data ($\leq$ 300 bit s^{-1}); voice-band data (300 bit s^{-1} to 16 kbit s^{-1}), and wideband data ($>$ 16 kbit s^{-1}). This type of classification by bit rate approximates to the transmission facilities required to support them. As an example of a system application for data services, Inmarsat-B uses the following:

- narrowband data via the SES TDMA 24 kbit s^{-1} channel with possibly 16 data bursts, each of 300 bit s^{-1}; modems would connect the CES to the PSTN;
- 9.6 kbit s^{-1} full duplex data on the 24 kbit s^{-1} data channel to permit packet data communications using, for example, the CCITT X.25 recommendation for the interface between Data Terminal Equipment (DTE) and Data-Circuit Terminating Equipment (DCE) for terminals operating in packet mode and connected to the PSDPN by dedicated circuits. This channel also supports CCITT Group-3 facsimile services; this service is also available in the SCPC voice channel using 2.4 kbit s^{-1} data rate and APC voice codecs. Wideband data will be supported as a later system with rates of at least 64 kbit s^{-1}.

Inmarsat-A provides wideband data facilities at a rate of 56 kbit s^{-1}. This high speed data transmission uses a voice channel on a dedicated frequency with a special modem. During transmission an Inmarsat-A earth station would need to have its EIRP increased by 2 dB (because of the use of a QPSK modulator instead of the FM voice modulator) to achieve the necessary quality for this service. The data stream is convolutionally encoded (rate 1/2) at the SES terminal and is decoded at the CES using a Viterbi soft decision decoder. Connection to the PSDPN is then possible in the same way as for the Inmarsat-B 9.6 kbit s^{-1} service.

Compandor circuit

Companding is a process whereby a voice signal is modified at the sending end in an attempt to improve the signal-to-noise ratio prior to modulation. The inverse of the process must be carried out at the receiver end in order to restore the original speech signal to its correct relative levels. The word *compandor* is a contraction of *com*pressor and ex*pandor*, which refer to the circuits at each end of the link. These circuits perform the task of modifying the speech signal. If the level of gain of the compressor and expandor circuits is controlled by the speech power at a syllabic rate, the compandor is referred to as a *syllabic compandor*.

In a typical installation the compressor circuit would amplify the high levels of microphone input less than the low levels. This would enable higher modulation levels to be achieved and improve the signal-to-noise ratio of the transmitted signal. A clipper circuit would limit the amplitude of any transients which are too fast for the compressor circuit to respond to. The compressed audio signal is then amplified, baseband limited by a filter (to give the required system baseband frequency range and remove band noise) and then used to frequency modulate the carrier.

At the receiver, after demodulation and filtering, the received audio signal is processed by the expandor circuit which removes the original compression by amplifying the high levels more than the low levels.

The amount of compression and expansion will depend on the characteristics of the circuit.

The expandor circuit has the added advantage that it will also heavily attenuate input noise making the receive channel sound quieter than would be the case without companding.

The advantage of using a compandor varies according to its use. Compandors used with an SCPC/FM system give an advantage of 15–20 dB.

4.6   Digital transmission

Introduction

Messages for transmission are often in digital form, e.g computer information, telex etc. Additionally, voice signals may be represented in digital form by a process of analogue-to-digital conversion. Regardless of whether the message source is digital or analogue, once it is in digital form it may be used to modulate a carrier for transmission. Additionally, if required, an encoder could be used to add redundant digits to the digital signal with the aim of improving the overall quality of the link. Demodulation and decoding is required at the receive end, together with digital-to-analogue conversion if required.

Advantages of digital transmission include:

- digital systems are less affected by noise, and other interference signals, when compared with analogue systems.
- digital systems are able to transmit both voice and data with equal efficiency.
- digital signals can be encoded for security, and improved quality of the link.
- digital systems are integrated more easily with terrestrial integrated service networks.

A digital SCPC system involves conversion of the analogue voice frequency signal to digital form using one of many coding techniques. Traditionally, for the PSTN, speech is considered to be formed by continuously varying waveforms in the audio frequency range 300–3400 Hz. Pulse code modulation (PCM) is a technique which samples the speech signal at regular intervals and interprets the sampled analogue level in digital form. The rate of sampling must be at least twice that of the highest baseband frequency to ensure the speech signal can be reconstructed later with acceptable quality.

PCM systems usually conform to CCITT recommendations with the sampling rate at 8 kHz, with each sample encoded into a 7-bit code giving 128 possible levels of representation. An eighth bit is used to represent the sign of the analogue signal. This results in a bit-rate for the PCM channel of $8 \times 8000 = 64$ kbit s^{-1}. Figure 4.24 shows a possible PCM coder/encoder circuit.

Figure 4.24 shows a quantizer encoder circuit, the function of which is to convert the pulse amplitude modulated (PAM) waveform to the uniform amplitude bit stream. The number of steps used (128) to produce the 7-bit code means that there will be a quantized amplitude, represented by the sample binary word, which may not exactly equal the quantity sampled. The difference between the actual sample and its quantized value gives rise to 'quantization noise' since, on reconstruction, the sample will be incorrect by the amount of sampling error. The effect of this could be reduced by increasing the number of sampling levels, and hence the number of bits that represent a

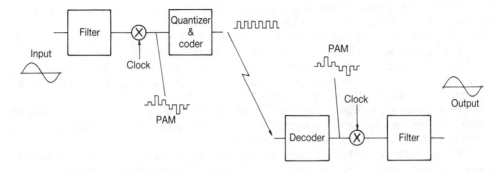

Fig. 4.24 Outline PCM coder/encoder circuit

sample. Since quantization noise is dependent on the step size between sampling levels, it follows that uniform quantization will give the same level of quantization noise for low amplitude signals as for those with large amplitudes. To improve this, non-linear quantization may be employed with smaller steps for low amplitude signals than for large amplitude signals. To achieve non-linear quantization a compandor circuit, similar in principle to that described earlier for analogue signals, could be utilized.

Other coding techniques include delta modulation (DM), differential pulse code modulation (DPCM) and adaptive variants such as APCM, ADM and ADPCM. In delta modulation, sampling of the voice signal is at a greater rate than for PCM and use is made of a single bit to track changes in input level from sample to sample. For adaptive DM, the quantizer step size is varied automatically according to the time-varying characteristics of the input signal. For DPCM, the difference between the sample, and estimates of it based on earlier samples, is quantized just as for PCM. At the receive end of the link the same predictions must be used as for the transmit end in order to add the same correction.

The coders described so far are all waveform encoders. A vocoder analyses speech in terms of a simplified model of speech production and sends the results of its analysis to the receiver for speech synthesization to reconstruct the original speech.

Inmarsat uses a system called adaptive predictive coding (APC) for the Inmarsat-B voice channel coding, and improved multi-band excitation (IMBE) speech coding for the Inmarsat-M voice channel. The IMBE system uses the vocoder principle.

4.7 Digital Modulation

Introduction

A carrier may have the form $V(t) = V_c\cos(\omega_c t + \phi)$, where V_c is the carrier amplitude, ω_c its angular frequency, t is time and ϕ the phase.

For digital modulation, the types available are amplitude-shift keying (ASK), frequency-shift keying (FSK) and phase-shift keying (PSK). Amplitude-shift keying can be accomplished simply by the on–off gating of a continuous carrier; frequency-shift keying by shifting the frequency of the carrier according to the binary transitions, and phase-shift keying by shifting the phase of the carrier according to the binary transitions. If the bit-stream at the modulator input has only two possible values (i.e. 0 and 1) then the system is binary. If the arrangement of bits gives more than two possible values (00,01 etc), then the system is referred to as *M*-ary where *M* is a value > 2.

Examples of *M*-ary modulation schemes include quadrature phase-shift keying (QPSK), offset QPSK (O-QPSK) and minimum shift keying (MSK).

A digital system is termed coherent if a local reference is available at the demodulator; that is, in phase with the transmitted carrier; otherwise the system is non-coherent. Similarly if, at the receiver, a periodic signal is available in synchronism with the transmitted digital sequence, the system is synchronous. If such a periodic signal is not required at the receiver the system is asynchronous.

The choice of a particular system will depend on many factors such as available power, bandwidth requirements, complexity of the equipment required and the effects of the transmission channel on the required signal.

Unlike analogue systems where system performance can be evaluated in terms of the signal-to-noise ratio (S/N), the performance for digital systems is measured by the bit-error rate (BER). The bit-error rate is defined as the number of bits transmitted that are received with errors, compared with the total number of bits transmitted. Typically the value of BER is specified by:

- average BER over a predetermined period of time;
- proportion of fixed-length time intervals which are either error free or undergo an error rate no worse than a specified value.

One of the most efficient modulation methods is PSK with coherent detection. This technique enables transmission of a constant envelope signal containing the required information as phase transitions—ideal for coherent detection. PSK is the only system used in the Inmarsat systems.

Phase-shift keying (PSK)

The simplest form of PSK is binary PSK or BPSK where the digital information modulates a sinusoidal carrier. This is shown in figure 4.25.

For phase modulation the phase of the carrier waveform varies according to changes in modulating amplitude. Generally:

$$\theta(t) = \omega_c t + \phi(t)$$

In BPSK, the frequency of the carrier remains constant while the phase shift is one of two fixed values. Consider figure 4.26 which shows the effect on carrier phase for a binary 1 and binary 0.

Figure 4.26 illustrates the effect on the carrier phase shown in the time domain. In the frequency domain the power spectral density of the modulated carrier varies as shown in figure 4.27 which illustrates that most of the energy is contained in the major lobe and, for a band-limited system, the bandwidth would be restricted to $2/T$. Since T represents the pulse duration, the bandwidth approximates to the bit-rate.

$\pm V_c \cos(\omega_c t + \theta_o)$

$\cos \omega_c t$

Fig. 4.25   Simple BPSK circuit

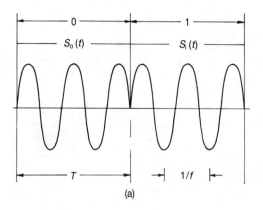

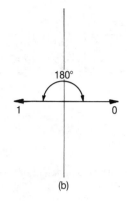

Fig. 4.26 180° phased shifted carrier

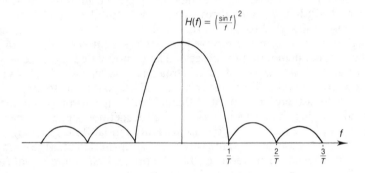

Fig. 4.27 Power lobes of a digitally modulated transmission

Restricting the bandwidth will result in the loss of some energy contained in the side lobes. However, the energy in those lobes removed in frequency from the main lobe is reduced with increasing frequency.

Incoherent detection for PSK is not feasible since it would be impossible to detect a 1 or 0. If a carrier of the same basic frequency as the transmitted signal is available at the receiver, synchronous demodulation is possible. For the arrangement shown in figure 4.28 the output will vary between two levels according to the phase of the input signal. However, with this arrangement it could be difficult to ensure which of the voltage output levels corresponded to the transmitted 1 or 0.

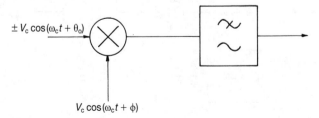

Fig. 4.28 A simple PSK synchronous demodulation circuit

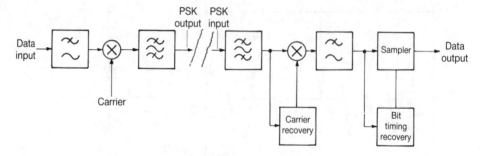

Fig. 4.29   Demodulation of PSK signals using a PLL

The principles outlined for the modulation/demodulation for a BPSK system can be expanded into the circuit of figure 4.29. On the modulator side a filter precedes the modulator in order to shape the modulation waveform and limit the bandwidth of the modulated signal. Post-modulation filtering is used to band-limit the transmitted signal. At the receiver the PSK signal goes through a pre-detector filter which limits the amount of channel noise allowed through to the modulator. A locally regenerated carrier component is extracted from the received carrier via a phase-locked loop (PLL) and assists in the product demodulation process. The demodulated signal passes through a low-pass filter and, via a second PLL, a bit-stream is produced which allows sampling of the signal at the pulse mid-point to reconstruct the original data stream.

As stated earlier, there may be a difficulty with this form of demodulation in successfully identifying the correct phase of the regenerated signal for demodulation. The ambiguity could be resolved by the use of a differential PSK system where, instead of the instantaneous phase determining which bit is transmitted, it is the change in phase which carries intelligence. This requires that the original baseband signal (unipolar or bipolar NRZ) waveform be recorded so as to register changes in phase with one logic level and no change in phase with the other logic level, i.e.

- if a digit changes (from 1 to 0 or vice versa), a 1 is transmitted;
- if no change occurs, a 0 is transmitted.

A basic demodulator arrangement for DPSK is shown in figure 4.30. The received waveform is delayed by one sampling period so that the multiplication is the product of a current sample and, what is in effect, a local oscillator input with the phase of the previous sample input. If both inputs to the multiplier have the same phase the demodulator output is positive; if the phases are different by 180° the demodulator output will be of negative sense.

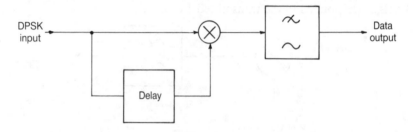

Fig. 4.30   A basic demodulator circuit for DPSK signals

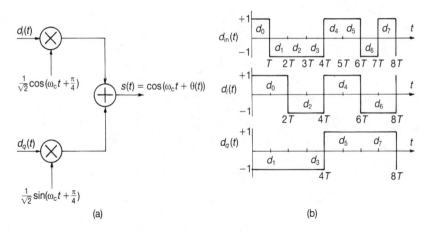

(a)   (b)

Fig. 4.31 A theoretical QPSK modulation system showing input and output signals

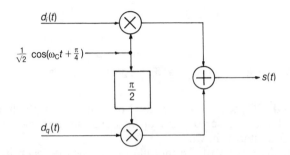

Fig. 4.32 A QPSK modulator circuit

Quadrature phase shift keying (QPSK)

A QPSK modulated signal can be formed by operating two BPSK modulators in quadrature. If the bit stream is split so that even numbered bits are routed to the in-phase stream and odd numbered bits to the quadrature-phase stream, then:

$$d_i(t) = d_0, d_2, d_4 \text{ etc}$$
$$d_q(t) = d_1, d_3, d_5 \text{ etc}$$

A possible bit stream input is shown in figure 4.31 together with its realization into in-phase and quadrature streams.

It can be seen from figure 4.31 that the two derived bit streams have half the bit rate of the input stream.

A QPSK modulator is shown in figure 4.32.

Pulse stream $d_i(t)$ is shown in figure 4.31 as a binary signal with $+1$ representing binary 1 and -1 representing binary 0. This stream will amplitude modulate the cosine function and will have the effect of shifting the phase of the function by 0 or π. This is BPSK. The pulse stream $d_q(t)$ will have the same effect on the sine function producing a BPSK waveform orthogonal to the cosine function. Summing the two orthogonal components results in the QPSK waveform.

Figure 4.33 shows the four phase states assumed for the QPSK signal. Each phase state depends on a pair of bits. As drawn, the *i*-channel bits operate on the vertical axis

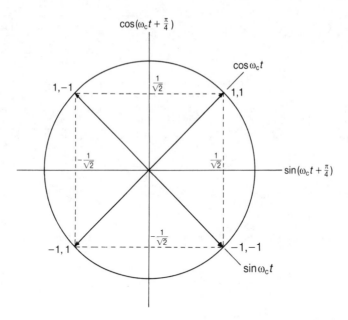

Fig. 4.33   The four phase states assumed for a QPSK signal

at phase states 90° and 270°, while the q-channel bits operate on the horizontal axis at phase states 0° and 180°. The vector sum of an i-channel and q-channel phase will produce each of the four states shown.

The phase state of the output of the modulator depends on each input channel so that the output state for each signal interval depends on a pair of bits. In this case the transmission rate depends on a pair of bits, and is measured in terms of symbols per second. For QPSK the power spectrum is the same as for BPSK but since the transmission rate in symbols per second is half the bit rate of BPSK, the bandwidth is halved. Figure 4.27 also serves for QPSK except that the time intervals shown are halved.

Because the i and q channels are orthogonal, coherent detection is possible in the receiver with each of the two BPSK signals being detected separately.

Offset QPSK (O-QPSK)

The basic concept of QPSK also applies to offset QPSK, the only difference being the timing of the two baseband signals. In QPSK, as shown in figure 4.31, each input pulse has a duration of T seconds and the odd/even streams have a duration of $2T$ seconds. All transitions are aligned in time as figure 4.31 shows.

For O-QPSK, the alignment of the odd/even streams is shifted so as to be offset by T seconds. The stream for $d_i(t)$ and $d_q(t)$ for figure 4.31 has been redrawn in figure 4.34 to show the effect.

For QPSK, carrier phase changes can occur only once every $2T$ seconds. Depending on the values of $d_i(t)$ and $d_q(t)$ the carrier phase at any instant will be as indicated by figure 4.33. At the next interval, if neither stream changes sign the carrier phase is unaltered; if one stream changes sign the carrier phase is shifted by $\pm 90°$; if both streams change sign, the carrier phase shift is 180°. If the QPSK signal is filtered to remove the

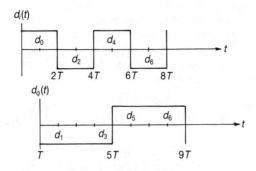

Fig. 4.34 The time shift produced by offset QPSK modulation

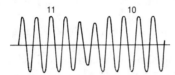

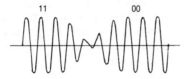

Fig. 4.35 QPSK signal envelopes showing changes in amplitude

spectral side lobes, the waveform that results no longer has a constant envelope. This effect is show in figure 4.35.

4.8 Coding

Introduction

Assuming that information to be transmitted is in digital form (where the information could be voice, telex or data) the channel through which the signal passes is likely to cause signal degradation. The noise and/or fading etc, experienced during transmission, could increase the probability of bit error at the received end. The data signal may be encoded in such a way as to reduce the likelihood of bit error. This process, known as coding, packages the bits that contain information with other bits before transmission. These other bits are known as redundant bits because they contain no information, but they can assist in the detection and correction of errors.

A relationship between a communication channel and the rate at which information can be transmitted over it has been established by C.E. Shannon. Basically, Shannon's rule states that if the information rate of a source is less than the channel capacity, there exist coding techniques which allow the transmission of the information with an arbitrarily small probability of error. The capacity (C) of the channel in bit s^{-1} with additive white Gaussian noise is given by:

$$C = B\log_2(1 + S/N)$$

where B is the channel bandwidth in Hz and S/N is the signal-to-noise power ratio within the bandwidth.

This relationship is known as the Shannon–Hartley Law.

The equation has been included because it does allow a quantitative value for channel capacity to be illustrated in terms of channel bandwidth and the S/N ratio within the bandwidth. For example, it would seem from the equation above that by increasing B, the channel capacity can be substantially increased. However, increasing B will increase N since the noise in the channel is proportional to B; thus if B increases, S/N decreases. Also, from the equation for channel capacity, it can be seen that if noise could be eliminated, the channel capacity would be infinite. Channels are noisy, however, and this gives a limiting value for channel capacity.

The main reason for using coding is to reduce the error rate, transmitter power level required etc, possibly at the expense of an increase in the required bandwidth. The main benefit of Shannon's equation is to show that coding will achieve that objective.

At the practical level, a system designer needs to strike a compromise between an acceptable error rate level and channel bandwidth, taking into account other factors imposed by the system—such as power levels available, system complexity etc.

One method of approaching Shannon's limit is to use a coding stream which allows for the detection, and correction, of any received errors. A possible coding system is shown in figure 4.36. The function of the encoder is to produce the required coding stream which is then used to modulate the carrier. The demodulator has the task of retrieving the transmitted coding stream from the received signal. There are two types of demodulation processes that can be used, namely:

- *hard decision*, where the demodulator produces its best attempt at representing the transmitted information;
- *soft decision*, where the demodulator additionally provides extra information on the validity of the hard decisions.

The decoder should use the redundancy of the coding stream to establish a valid received sequence.

The types of errors associated with digital transmission are:

- *random*, where there is no correlation between the symbols in error;
- *burst*, where a group of consecutive symbols are likely to be in error.

Fig. 4.36   A signal channel showing coding and decoding arrangements

The types of codes chosen may be selected on the basis of how well they deal with each of these errors, or as a combination of both of them.

Block codes

A coding stream could use what is known as a block code. This type of code uses input data divided into k information symbols, together with redundant check symbols, to produce a code word of n symbols. An encoder operating in this way would produce a (n, k), or rate k/n, block code.

For a digital system the bit is the basic element and can have the value of binary 0 or binary 1. Bits could be grouped to provide the information that is to be transmitted. For two equal length binary code words there is a distance between them where distance is defined as the number of bit positions in which the two words differ. As an example, if the first code word (s_i) = 0010011, the second code word (s_j) = 1101011, then the modulo-2 addition of s_i and s_j gives 1111000 which shows there are 4 bits different out of the total of seven bits, i.e. the distance between the two codes (d_{ij}) = 4.

Cyclic codes

A type of parity check code, called cyclic code, is implemented simply by the use of electronic circuits known as shift registers. Any cyclic codeword can produce another valid codeword by giving an end-to-end shift of one digit, i.e. if a codeword is given by:

$$u_0, u_1, u_2 \ldots u_{n-2}, u_{n-1}$$

then

$$u_{n-1}, u_0, u_1, u_2 \ldots u_{n-2}$$

is another valid codeword.

An (n, k) cyclic code can be generated using a $(n-k)$ shift register and a process of modulo-2 division. A binary number of length k which consists of n information bits is followed by $(k-n)$ binary zeros and divided by a binary number of length $(k-n+1)$ bits. Once the process of division takes place, the remainder consists of $(k-n)$ bits which are then added to the n information bits to produce the unique codeword. As an example, consider the $(7, 4)$ cyclic code and code word 1000101 which has information bits 1000 and parity bits 101. A shift would give a new codeword 0001011 which corresponds to the information bits 0001. The divisor for this unique set of codewords is 1011, i.e.

```
1 0 1 1 )1 0 0 0 0 0 0
         1 0 1 1
         1 1 0 0
         1 0 1 1
           1 1 1 0
           1 0 1 1
             1 0 1
```

In this example the information bits were 1000 and were added to three zeros to give 1000000 which, when divided by 1011, gave a remainder of 101. The remainder replaces the three zeros at the end of the word 1000000 to give 1000101 etc.

Addition and subtraction in modulo-2 arithmetic are identical processes and the codeword must be a perfect multiple of the divisor. At the receive end the codeword is divided by the same divisor as used at the transmitting end in order to recover the information. A remainder of zero will indicate that the codeword received is a valid member of the set.

Bose–Chadhuri–Hocquenghem (BCH) codes

BCH codes are a class of cyclic codes with a large range of block lengths, code rates, alphabets and an error-correcting capability. BCH codes have been found to be superior in performance to all other codes of similar block length and code rate. Most commonly used BCH codes have a codeword block length of $n = 2^m - 1$, where $m = 3, 4, \ldots$. The Inmarsat-A system uses BCH codes for its TDM channels and for the request channel burst.

Convolutional codes

This type of code is not a block code. Instead of parity bits being added to form a block of data, the parity bits are calculated over a longer span of bits to form a continuing message sequence. A convolutional code can be described in terms of integers n and k. Integer n is a measure of the number of coded bits that go into making the sequence while k is a representation of the span of bits forming the message. The factor k is known as the constraint length.

4.8   Decoding

Block Codes

The simplest means of decoding block codes is by a method of correlation whereby the decoder makes a comparison between the received codeword and all permissable codewords, selecting that word which gives the nearest match. Decoding of such codes will also depend on whether error detection or error correction is required. Decoders of block codes generally cannot use soft decision outputs from the demodulator, unlike the decoders for convolutional codes.

Convolutional codes

The complete transmission loop requires a convolutional encoder followed by modu-lation, transmission channel, demodulator and decoder. The effect of the transmission channel on the signal, and the probability of detection of a 1 or 0 in the presence of Gaussian noise, has been discussed previously. Output from the demodulator can be configured to give a hard decision regarding whether the signal is 1 or 0. The process of decoding then depends on the two state inputs it receives. An alternative demodulator configuration allows quantization of the predicted level which gives the decoder more information regarding the probable state of the demodulator output. For example, if 3 bit ($2^3 = 8$ levels) quantization occurs then 0 0 0 would suggest a firm valuation of the level received as a 0. On the other hand 0 0 1 suggests the 0 is received close to the threshold and this valuation as a 0 is made with less certainty. The reason for quanti-zation is to provide the decoder with more information in order to recover correctly the transmitted information with better error performance probability.

4.9   Error correction

Forward Error Correction (FEC)

Forward Error Correction (FEC) coding, which is the result of convolutional coding, is used in Inmarsat systems for some voice and telex channels and signalling channels. For

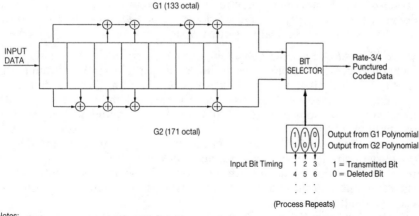

Notes:
– The first bit in each transmission frame is the output from the G1 polynomial
– All bits are transmitted for the rate -1/2 code

Fig. 4.37   Encoder logic including punctured operation

example, Inmarsat-B uses a convolutional encoder of constraint length 7 and an 8-level soft-decision Viterbi decoder. The coding rate is either 3/4 or 1/2, see figure 4.12. For voice channels the rate 3/4 code is used and is derived by puncturing the rate 1/2, $k = 7$ convolutional code. The generator polynomials for the rate 1/2, $k = 7$ convolutional code are:

$$G_1 = 1 + X^2 + X^3 + X^5 + X^6$$
$$G_2 = 1 + X + X^2 + X^3 + X^6$$

The encoder logic, including the punctured operation is shown in figure 4.37.

In the 1/2 rate code, the relationship between three input bits and the six output bits is given by:

Input bit time	1	2	3
Output sequence	$G_1 G_2$	$G_1 G_2$	$G_1 G_2$

The rate 3/4 coded sequence is obtained by eliminating 2 bits from each block of six output bits from the rate 1/2 convolutional encoder as follows :

Input bit time	1	2	3
Output sequence	$G_1 G_2$	G_1	G_2

The receiving end is informed of the sequence of the four output bits, which make up the punctured code block, by arranging for the first boundary of the punctured code in each 80 ms transmission frame to coincide with the first bit that follows the last bit of the unique word or framing bit pattern.

At the beginning of each voice activated burst the shift register of the FEC encoder is set to the all-zero state. The Viterbi decoder at the receive end assumes this initial state at the start of each burst.

Automatic Request Repeat (ARQ)

Forward Error Correction (FEC) requires only a one-way transmission link since the message contains parity bits used for detection and correction of errors. Automatic

Request Repeat on the other hand requires a two-way link since a receiver, detecting an error, does not attempt to correct it but simply requests the transmitter to retransmit the message. ARQ systems are basically of three types.

- *Stop and Wait ARQ*. Each message block is transmitted and the transmitter waits for an acknowledgement before transmitting a further block. A half-duplex link is required (i.e. transmission on the link is possible in both directions but not at the same time). If a message block is correctly received the next message block is transmitted; if the block received is in error the transmitter will be instructed to retransmit that block.
- *Continuous ARQ with Repeat*. The transmitter sends continuously and the receiver acknowledges continuously. Any message block not correctly received causes the transmitter to return to the block in question and re-commence continuous transmission from there. A full-duplex link is necessary (i.e. the ability to transmit in both directions simultaneously).
- *Continuous ARQ with Selective Repeat*. In this arrangement only the block received in error is retransmitted and the transmitter continues from where it left off instead of repeating any subsequent correctly received messages. Again a full-duplex link is necessary.

A major advantage of ARQ compared with FEC is that decoding equipment for error correction can be simpler and the redundancy in the total message stream is less. ARQ efficiency is good for low error ratios but for high error ratios requiring retransmission of a large number of message blocks the system becomes inefficient. A disadvantage of ARQ is the variability of the delays experienced from end-to-end of the link and the possible requirement for large data stores for incoming data blocks.

Continuous ARQ may be used in conjunction with FEC to provide a hybrid system. Such an arrangement could be used to provide feedback information to the transmitter regarding slow variations, such as fading. The Inmarsat-C system uses packets of data and every packet transmitted contains a 16-bit checksum field. The receiver completes an expected checksum for each packet and compares this with the actual packet received in order to verify that the packet has been correctly received. ARQ is used if the packet received is in error.

Pseudo-noise

A pseudo-noise (PN) generator will generate a set of cyclic codes with good distance properties. The name of the sequence is given because the sequence, although deterministic, appears to have the properties of sampled white noise. A PN sequence is easily generated using shift registers and has a correlation function that is highly peaked for zero delay and approximates to zero for other delays. The PN sequence, being deterministic, is useful for synchronization purposes between a transmitter and receiver.

Some Inmarsat systems use a scrambler circuit before FEC encoding and a descrambler at the receive end following FEC decoding (see figure 4.17). For Inmarsat-B and Inmarsat-M systems, for example, the scrambler/descrambler circuits are PN generators using 15 stages.

The scrambler/descrambler circuits are clocked at the rate of one shift per information bit. The first bit into the scrambler at the beginning of a frame is modulo-2 added with the output of the scrambler shift generator corresponding to the initial state scrambling vector. The initial state of the shift register is set at the beginning of a burst and a frame.

Considering the Inmarsat-M system, the initial state of the scrambler shift register at the LES (for the SCPC channel operating in voice mode) is sent to the LES by the MES at the start of a call as part of the call set-up sequence. The MES chooses any initial state (except all zeros) on a random basis for each call and signals this 'scrambling vector' message (1 0 0 0 1 1 0 1 or 8D in hexadecimal form) for implementation at the LES with the least significant bit (LSB) in shift register #1 and the most significant bit (MSB) in shift register #15 of the scrambler. The MES simultaneously sets the descrambler shift registers with the same scrambling vector. For MES to LES channels, a fixed initial state default value of 1 1 0 1 0 0 1 0 1 0 1 1 0 0 1 (6 9 5 9 in hexadecimal form) is used in the MES scrambler and LES descrambler

Interleaving

Channel impairments such as fading and multipath, which exist on a satellite channel, cause statistical dependence among successive symbol transmissions, i.e. errors tend to occur in bursts. Fading may be a slow variation compared with the time occupied per symbol.

Multipath is caused by signals arriving at the receiver via two or more paths of different lengths; this results in phase differences between received signals and causes distortion of the received signal. Since most block and convolutional codes are designed to resist random independent errors, the effects caused by fading and multipath would render the system inoperative. The process of interleaving causes bursts of errors to be spread out in time and these can thus be decoded as if they were random errors. The interleaver acts as a sort of time-division multiplexer with the symbols of one codeword mixed in time with the symbols of other codewords.

Suppose a codeword is made up of seven symbols and there are seven such codewords in a block (for block codes) or constraint length (for convolutional codes). Each codeword is made up of symbols An_1 to An_7 where A can have the value 1 to 7 i.e.:

$$n_1,n_2,n_3,n_4,n_5,n_6,n_7,2n_1,2n_2,2n_3,2n_4,2n_5, \ldots 7n_5,7n_6,7n_7$$

If the symbols were interleaved such that the pattern transmitted is of the form:

$$n_1,2n_1,3n_1,4n_1,5n_1,6n_1,7n_1,n_2,2n_2,3n_2 \ldots 4n_7,5n_7,6n_7,7n_7$$

then if an error burst occurs that lasts for seven symbol periods, only one symbol from each of the original codes would be affected. Once received and de-interleaved, the original codewords would appear, each with a single bit error. If each codeword possesses a single error-correcting capacity the words should be decoded satisfactorily. Without interleaving, the seven symbol burst could have destroyed one or two of the original codewords.

For block interleaving, the coded symbols are received from the encoder in blocks. The interleaver permutes the symbols and sends the rearranged symbols to the modulator. A usual form of permutation is the use of an array (matrix) where the blocks are entered by filling the columns of the array and, once the array is full, the symbols are fed to the modulator one row at a time. At the receiver, symbols would be entered into a similar array by rows and taken out to the decoder on a column by column basis. Figure 4.38 illustrates an array of dimension 7×7 filled with seven codewords each of seven symbol length. Codeword 1 (n_1 to n_7) fills column 1, codeword 2 ($2n_1$ to $2n_7$) fills column 2 and so on. The output from the interleaver would be ($n_1,2n_1,3n_1,4n_1,5n_1,6n_1,7n_1$) followed by: ($n_2,2n_2,3n_2,4n_2,5n_2,6n_2,7n_2$) etc.

The memory requirement at the receiver would be for $X.Y$ symbols (where $X =$ the number of columns, $Y =$ the number of rows). Typically, however, since the array needs to be mostly filled before it can be emptied, a second array of $X.Y$ symbols is additionally employed so that it can be filling while the first array is being emptied. The use of interleaving with a single error correcting code requires that the number of

7 X 7 DIMENSIONAL ARRAY						
n_1	$2n_1$	$3n_1$	$4n_1$	$5n_1$	$6n_1$	$7n_1$
n_2	$2n_2$	$3n_2$	$4n_2$	$5n_2$	$6n_2$	$7n_2$
n_3	$2n_3$	$3n_3$	$4n_3$	$5n_3$	$6n_3$	$7n_3$
n_4	$2n_4$	$3n_4$	$4n_4$	$5n_4$	$6n_4$	$7n_4$
n_5	$2n_5$	$3n_5$	$4n_5$	$5n_5$	$6n_5$	$7n_5$
n_6	$2n_6$	$3n_6$	$4n_6$	$5n_6$	$6n_6$	$7n_6$
n_7	$2n_7$	$3n_7$	$4n_7$	$5n_7$	$6n_7$	$7n_7$

Fig. 4.38   An array of dimension 7 x 7 filled with seven codewords each of seven symbol length

columns (X) must exceed the expected burst length. The number of rows (Y) depends on the coding scheme used. For block codes the value of Y should be greater than the block length while for convolutional codes Y should be greater than the constraint length. If this requirement is implemented then a burst of length X will cause no more than a single error in the block codeword (for block codes) or a single error in the constraint length (for convolutional codes).

The Inmarsat-C system uses the interleaving process where data is transmitted in blocks using 640 bytes. A 1/2 rate convolutional encoder produces $640 \times 8 \times 2$ symbols (10240 symbols) which are then passed to an interleaver.

5
The Inmarsat organization

5.1 Introduction

This chapter looks in detail at the *International Maritime Satellite Organization (Inmarsat)* which, through the establishment and use of several geostationary satellites, covering four ocean regions, provides global communications for mobile users. Whilst Inmarsat was originally established to serve the communication needs of the international maritime industry, the organization's sphere of influence has now been extended to aeronautical, vehicular and land-based mobile systems. The Inmarsat communications system is the product of extraordinary co-operation between commercial companies, government authorities and telecommunications organizations around the world. These bodies are all part of the Inmarsat system partnership and they include the following.

Inmarsat signatories

Each member Government nominates an organization to be its Inmarsat signatory. That body then becomes its financial shareholder in Inmarsat, takes part in the Inmarsat decision making processes and, usually, provides Inmarsat services for its country. These organizations are usually national telecommunications authorities, although some are Government departments or private companies.

Land earth station operators

Land earth station (LES) is the generic term used to describe the station providing the fixed part of the communications link to and from the satellite. The term *coast earth station* (CES) is used for maritime operations and *ground earth station* (GES) for aeronautical operations.

LES operators are invariably Inmarsat signatories, but not always. These organizations own and operate the LES which provides the interface between the Inmarsat satellite system and national/international fixed telecommunications networks. Each LES operator establishes its own range of mobile service offerings and sets its own user charges. Mobile users can choose which LES operator, within their ocean region, they will access for a particular service.

National communications authorities

Communications services from the point of origin to an LES are provided through national telecommunications organizations, which set their own charges and arrange routeing through a convenient LES.

Equipment manufacturers

Inmarsat does not manufacture equipment. Instead it devises the standards to be met by communications terminals using its satellites, and it occasionally sponsors the development of prototypes. The equipment required for customers to use the Inmarsat system is provided by major international electronics manufacturers, distributed and serviced through their world-wide dealer networks.

Value added service providers

These organizations, generally private companies, offer enhanced services, databases and management systems, via the Inmarsat system. Such services may include electronic mail etc.

The Inmarsat organization

The Inmarsat organization is a co-operative, commercially orientated enterprise with currently 64 member countries. Inmarsat is headquartered in London, UK. Its prime task is to establish, maintain and operate the satellite system in order to provide global mobile communications.

Inmarsat was created in 1979 in order to provide global communications for the maritime industry. On 1 February 1982 Inmarsat started to provide these services when it took over and expanded the satellite communications system established in 1976 by the US Marisat consortium. The scale of system expansion since 1982 has been breathtaking. Newer and bigger satellites have increased channel capacity by thousands of percentage points. Mobile user terminal fittings of Inmarsat-A have increased from a mere handful, in 1982, to over 16,000 in the period of one decade. A figure which will increase further as the Inmarsat-B system replaces Inmarsat-A. Inmarsat-C, a system handling data only, has been established, and now has nearly 3000 mobile terminals accessing the system. It is predicted that there will be some 600,000 Inmarsat-M MESs in use by the year 2005.

In 1992 there were 22 LESs providing global services on the Inmarsat-A system, and a further 11 are planned for the near future. Eight LESs are providing services in the Inmarsat-C system, with a further 13 planned for the near future. There are ten ground earth stations (GES) providing aeronautical services, with numerous others being planned for the near future.

Figure 5.1 provides information on the status of CES fittings internationally.

Two new technically advanced communications systems, Inmarsat-B and Inmarsat-M started in 1993 to expand further the satellite mobile communications service.

Inmarsat is the only provider of global communication services for mobile users. The telecommunications industry never remains static and Inmarsat is no exception. Additions to the system over the next decade into the new millennium will include some of the following:

● increased channel capacity by the use of Inmarsat–3 (INM3) satellites;
● the expansion of the new systems Inmarsat-B and Inmarsat-M;
● the availability of position determination and navigation applications using Inmarsat satellites;
● the introduction of the world's first global paging service;
● the introduction of a world-wide pocket-sized telephone service.

Whilst it is unlikely that all of the above innovations will be fully available by the year 2000, it is likely that most of them will be in the final stages of development.

COAST EARTH STATIONS

Inmarsat-A

Country	Location	Operator	Coverage Region	Access (Octal)	Code (Decimal)
Australia	Gnangara	IDB Comms Group Inc	IOR	13	11
Australia	Perth	Telstra	IOR/POR	02	02
Brazil	Tangua	EMBRATEL	AORE	14	12
China	Beijing	Beijing Marine Coms & Nav Co	POR/IOR	11	09
Denmark, Finland,	Eik	Norwegian Telecom	IOR/AORE		
Iceland, Norway, Sweden			/AORW	04	04
Egypt	Maadi	National Telecoms Organisation	AORE	03	03
France	Pleumeur Bodou	France Telecom	AORW/AORE	11	09
Germany	Raisting	Fernmeldetechnisches Zentralamt	AORE	15	13
Greece	Thermopylae	OTE SA	IOR	05	05
India	Aarvi	Videsh Sanchar Nigam Ltd	IOR	06	06
Iran	Boumehen	Telecomm Co of Iran	IOR	14	12
Italy	Fucino	Telespazio	AORE	05	05
Japan	Yamaguchi < + >	Kokusai Denshin Denwa	POR/IOR	03	03
Korea,Republic	Kumsan	Korea Telecom Authority	POR	04	04
Netherlands	Burum	PTT Nederland NV	AORE/IOR	12	10
Poland	Psary	Ministry Transp/Marit Economy	AORE/IOR	16	14
Russia	Nakhodka	Morsviazsputnik	POR	12	10
Saudi Arabia	Jeddah	Ministry of PTT	IOR	15	13
Singapore	Sentosa	Singapore Telecom	POR	10	08
Turkey	Anatolia	Comsat Corporation	IOR	01	01
Turkey	Ata	General Directorate of PTT	IOR/AORE	10	08
UK	Goonhilly	BT International	AORW/AORE	02	02
Ukraine	Odessa	Morsviazsputnik	AORE/IOR	07	07
USA	Niles Canyon	IDB Comms Group Inc	POR/AORW	13–1	11–1
USA	Santa Paula	Comsat Corporation	POR	01	01
USA	Southbury	Comsat Corporation	AORW/AORE	01	01
USA	Staten Island	IDB Comms Group Inc	AORE	13–1	11–1

Inmarsat-C

Country	Location	Operator	Coverage	Code
Australia	Perth	Telstra	IOR/POR	302/202
Brazil	Tangua	EMBRATEL	AORE	114
Denmark, Finland	Blaavand	Telecom Denmark	AORE	131
Iceland, Norway, Sweden				
Denmark, Finland,	Eik	Norwegian Telecom	IOR	304
Iceland, Norway, Sweden				
France	Pleumeur Bodou	France Telecom	AORW/AORE	011/111
Germany	Raisting	Fernmeldetechnisches Zentralamt	AORE	115
Greece	Thermopylae	OTE SA	IOR	305
Netherlands	Burum	PTT Nederland NV	AORE/IOR	112/312
Poland	Psary	Ministry Transp/Marit Economy	AORE	116
Singapore	Sentosa	Singapore Telecom	POR	210
UK	Goonhilly	BT International	AORW/AORE	002/102
USA	Santa Paula	Comsat Corporation	POR	201
USA	Southbury	Comsat Corporation	AORW/AORE	001/101

Fig. 5.1 Inmarsat coast earth stations. (Courtesy *Ocean Voice*)

Inmarsat-M/B

Country	Location	Operator	Coverage	System
Australia	Perth	Telstra	POR/IOR	M/M
Japan	Yamaguchi	Kokusai Denshin Denwa	POR/IOR	M + B/M + B
UK	Goonhilly	British Telecom International	AORW/AORE	M/M
USA	Santa Paula	Comsat Mobile Communications	POR	M + B
USA	Southbury	Comsat Mobile Communications	AORW/AORE	M + B/M + B

Planned

Country	Location	Operator	Coverage	System	Status
Argentina	Balcarce	Comision Nac de Telecom	AORE	A	1993/95
China	Beijing	Beijing Marine Coms & Nav Co	POR/IOR	C	1993
Cuba	N/A	Ministry of Communications	AOR	A	1993/95
Denmark, Finland, Iceland, Norway, Sweden	Eik	Norwegian Telecom	AORW	A	1993
France	Aussaguel	France Telecom	AORW/AORE/IOR	M/B	1993
Germany	Raisting	Fernmeldetechnisches Zentralamt	IOR	A/C/M	1994
Greece	Thermopylae	OTE SA	AORE/IOR	M/B	1995
India	Aarvi	Videsh Sanchar Nigam Ltd	IOR	C	1993
Iran	Boumehen	Telecomm Co of Iran	IOR	C	1993
Italy	Fucino	Telespazio	AORE	C/M	1993/95
Korea,Republic	Kumsan	Korea Telecom Authority	IOR/POR	A/C	1993
Kuwait	Umm-al-Aish	Ministry of Communications	AORE	A	1994
Netherlands	Burum	PTT Nederland NC	AORE/IOR	M/B	1994
Poland	Psary	Ministry Transp/Marit Economy	IOR	C	Planned
Portugal	Lisbon	Comp. Portuguesa Radio Marconi	AORE	C	1993
Russia	Nakhodka	Morsviazsputnik	POR	C	1994
Saudi Arabia	Jeddah	Ministry of PTT	AORE/IOR	A/C/M/B	Planned
Singapore	Sentosa	Singapore Telecom	IOR/POR	C/M/B	1993/94
Spain	Buitrago	Telefonica de Espana SA	AORE	A/C	1993/95
Turkey	Ata	General Directorate of PTT	IOR	C	1993
UK	Goonhilly	BT International	AORW/AORE	B	1993
Ukraine	Odessa	Morsviazsputnik	AORE/IOR	C	1994
USA	Niles Canyon	IDB Comms Group Inc	AORW	M/B	1995
USA	Staten Island	IDB Comms Group Inc	AORE	M/B	1995

Key

AORE	Atlantic Ocean Region East		IOR	Indian Ocean Region
AORW	Atlantic Ocean Region West		POR	Pacific Ocean Region

Fig. 5.1   (cont.)

The space segment

Satellites must inevitably form the nucleus of any truly global communications system. As the reader will appreciate from the previous chapter, a satellite is an extremely expensive and complex piece of hardware. It is violently hurtled into orbit, and expected to survive in a hostile environment subject to extreme temperature changes and high levels of radiation. However, space technology has progressed rapidly over the past three decades to the point where satellites can be relied upon to operate virtually faultlessly and provide high quality communications throughout their operational life. The Inmarsat organization bases its earth coverage on a constellation of four prime

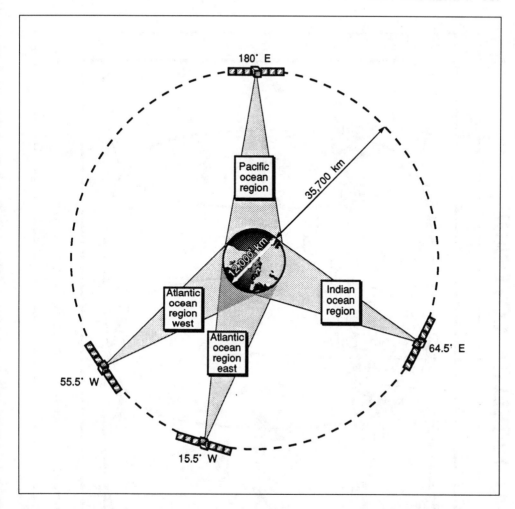

Fig. 5.2   View of the satellites in geostationary orbit above the four ocean regions. (Courtesy Inmarsat)

geostationary satellites covering four ocean regions. Two in geosynchronous orbit over the Atlantic Ocean region east and west (AORW and AORE), one positioned over the Indian Ocean region (IOR) and one over the Pacific Ocean region (POR). Several other satellites are maintained in orbit as back-up spares.

Figure 5.3 shows the footprints projected onto the surface of the earth from the four prime Inmarsat geostationary satellites currently in use. It should be noted that the recommended limit of latitudinal coverage is within the area between 75° north and south. However, as a large percentage of the earth's mobile communication requirements lie within this area, the system is considered to possess a global coverage pattern. The original Inmarsat system was based on a three satellite constellation of AOR, IOR and POR. With the rapid growth of maritime satellite communications traffic in the 1980s and in order to close a small gap in coverage which existed it was decided that a second satellite would be needed in the Atlantic Ocean region. Consequently, the space segment was expanded to be four ocean regions each with prime satellite coverage. It

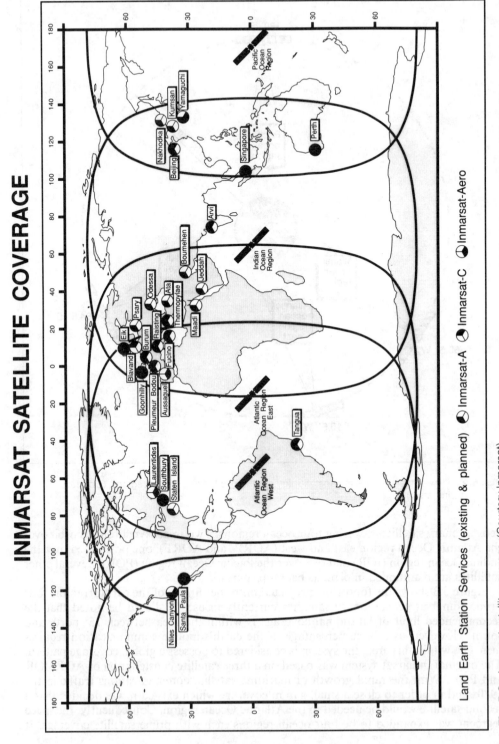

Fig. 5.3   Inmarsat satellite coverage. (Courtesy Inmarsat)

SATELLITE STATUS 1993				
Ocean Region	Spacecraft	Location	Launch Date	Status
AORW	INM2–F4	55.5W	1992	Operational
	Marecs–B2	58.0W	1984	Spare
AORE	INM2–F2	15.5W	1991	Operational
	Intelsat V–MCS B	18.5W	1983	Spare
	Marisat–F1	106.0W	1976	Spare
IOR	INM2–F1	64.5E	1990	Operational
	Intelsat V–MCS A	66.0E	1982	Spare
	Marisat–F2	72.6E	1976	Spare
POR	INM2–F3	179.0E	1992	Operational
	Intelsat V–MCS D	180.1E	1984	Spare
	Marisat–F3	182.5E	1976	Spare

Fig. 5.4   Satellite status 1993. (Courtesy *Ocean Voice*)

should be noted that there are considerably more than four Inmarsat operated satellites in orbit. It would be folly indeed to rely upon a single operational satellite in each ocean region.

In February 1982 Inmarsat took over and expanded the communications system started in 1976 by the US-based COMSAT group, Inmarsat's US signatory. The Marisat system was formed by three satellites interlocking three ocean regions. Each satellite was able to provide a mere ten communication channels, a fact which, because of the rapid growth of ship earth station installations, caused queuing to occur, particularly in the AOR. Three Marisat satellites are, in fact, still in orbit providing back-up facilities as required.

The constant battle to keep the number of available channels ahead of demand led to Inmarsat's decision to lease a 30 channel capacity package, known as the Maritime

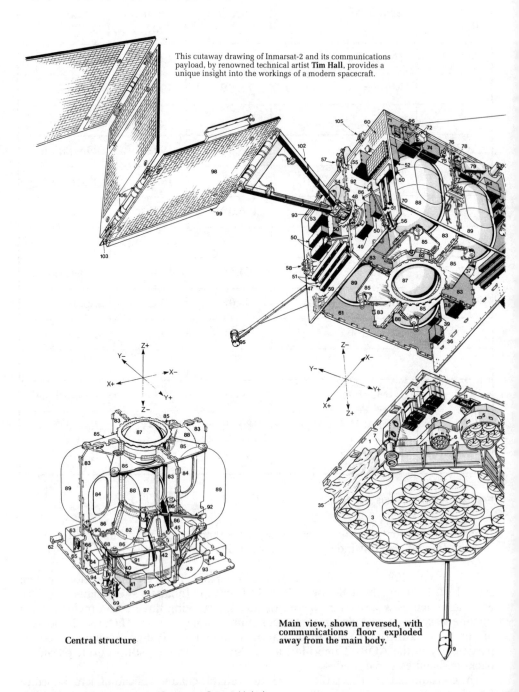

This cutaway drawing of Inmarsat-2 and its communications payload, by renowned technical artist **Tim Hall**, provides a unique insight into the workings of a modern spacecraft.

Central structure

Main view, shown reversed, with communications floor exploded away from the main body.

Fig. 5.5   Inside Inmarsat 2. (Courtesy *Ocean Voice*)

KEY - INMARSAT-2 SATELLITE
COMMUNICATIONS FLOOR
1   Floor structure of aluminium skins and honeycomb core
2   Antenna farm platform (aluminium skin and honeycomb core)
3   L-band transmit antenna
4   L-band receive antenna
5   C-band transmit antenna
6   C-band receive antenna
7   Tracking, telemetry and command (TTC) fill-in antenna
8   Fixed omni antenna
9   TT&C omni antenna
10   Infra-red 2-axes earth sensors
11   Isolators, LP filter and test coupler
12   Nutation sensor package
13   Payload interface unit
14   Switches
15   Power combiner

16   L-band output band pass filter (BPF)
17   Power monitor
18   Low pass (LP) filters
19   Isolators
20   Sum port monitor
21   Struts (10) connecting communication floor to antenna farm platform
22   Directional filters
23   L-band receiver
24   Channelization assemblies
25   C-band driver linearizer
26   Upconverter ALC
27   Channel filter
28   C-band output band pass filter (BPF)
29   C-band input filter
30   BPF receivers
31   Signal divider
32   Switch driver
33   Command attenuator (ALC)
34   L-band pass filter (BPF)
35   Multi-layer insulation (MLI)
COMMUNICATIONS PAYLOAD
Y+ WALL
36   Wall structure (aluminium with honeycomb core)
37   Solar array drive mechanism (SADM)
38   Driver amplifier linearizer L-band (PSDAL's)
39   TT&C C-band transponders
40   Battery
41   Central interface unit (CIU)
42   Array switching regulator (ASR)
43   Fixed momentum wheel (FMW)
44   Battery control and interface unit (BCIU)
45   Electronic power conditioners L-band (EPC's) (3) - #4 (1), #5 (2)
46   Travelling wave tubes L-band (TWT's)
Y- WALL
47   Wall structure (aluminium with honeycomb core)
48   SADM
49   PSDAL's
50   EPC's (5) - #1 (2), #2 (2), #3 (1) - (C-band 2 off, L-band 3 off)
51   TWT's (C-band 2 off, L-band 3 off)
52   Battery
53   BCIU
54   FMW
55   Control law electronics (CLE)
56   Thruster module 1A/1B
57   Thruster module 2A/2B
58   Thruster module 3A/3B
59   Thruster module 6A/6B
60   Thermal control mirrors

SPACECRAFT SYSTEMS
X- WALL
61   Wall structure (aluminium with honeycomb core) in two sections, upper access panels & lower service module
62   Sun acquisiton sensor (SAS)
63   Actuator drive electronics (ADE)
64   Battery discharge regulator (BDR)
65   Pyro safe and arm connectors
66   Battery safety connector
67   TT&C Connector
68   Power subsystem (PSS) connectors
69   Thruster module 5A/5B
X- WALL
70   Wall structure (aluminium with honeycomb core) with two sections, upper access panels & lower service module
71   Attitude & orbit control system connector (AOCS)
72   Earth sun sensor (ESS)
73   Gyro
74   BDR's
75   Battery safety connector
76   Clearance hole for relief valve
77   Pyro safe/arm connectors
78   Automatic hold up circuit (AHC)
79   Battery reconditioning unit
80   SAS
81   Thruster module 4A/4B
CENTRAL STRUCTURE AND GENERAL
82   Central structure of carbon fibre reinforced plastic
83   Shearwalls - aluminium with honeycomb core
84   Cutouts for tanks in +X and -X shearwalls
85   Upper tank support panels (aluminium with honeycomb core)
86   Lower tank support panels
87   Pressurant tanks - spherical (2) (Helium)
88   Fuel tanks (2) (Mono methyl hydrazine)
89   Oxidant tanks (2) (Nitrogen tetoxide)
90   Tank support struts (8)
91   Apogee boost motor
92   Struts (6) central structure to +Y and =Y panels
93   Closure panels (2), -Z
94   Z telemetry fill-in horn
95   Stability booms (2)
96   Stability boom hold-down point (2)
97   Electrical umbilical connectors (2)
SOLAR ARRAY
98   Solar panels - single sided panels (array is on a 5deg eccentric axis to satellite body)
99   Thermal fins
100   Attitude and orbit control subsystem (AOCS) flap
101   Hinges
102   Yoke
103   Solar array sun sensors (SASS)
104   Hold down points on solar panels
105   Hold down points (4) two not shown, on -Y and +Y panels
106   Flexible interpanel harness

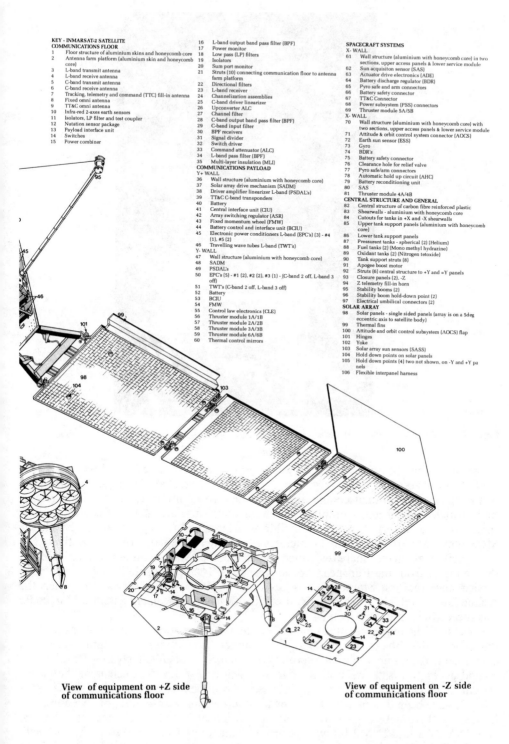

View of equipment on +Z side of communications floor

View of equipment on -Z side of communications floor

Fig. 5.6   INM2 satellite showing the antennae array. (Courtesy *Ocean Voice*)

Communications Subsystem (MCS), on several Intelsat-V satellites. These satellites, made by Ford Aerospace, are large capacity vehicles which handle a high volume of the earth's fixed international communications links. In addition, each satellite carries a MCS which is dedicated to mobile communications use and leased from Intelsat by Inmarsat. The Intelsat V-MCS-A satellite was launched in September 1982 with the MCS-B and MCS-C and MCS-D satellites following in 1983 and 1984 respectively.

Capacity was further increased when the 40 channel Marecs satellites were leased in their entirety from the European Space Agency (ESA). These satellites have provided service since the first Marecs A was launched in 1982. Marecs B was lost when the Ariane launch vehicle failed shortly after lift off. The replacement satellite Marecs B2 was successfully launched in 1984.

Because of the unprecedented demand on the system, it soon became evident that, by the end of the 1980s, the demand for channels would outstrip supply. Inmarsat commissioned a number of manufacturers to produce what effectively may be called the second generation satellite system, four of which now form the nucleus of the global system. The second generation of Inmarsat developed satellites, called Inmarsat-2 (INM2) have been designed and built by an international consortium of six companies headed by British Aerospace to provide a theoretical 250 channel capacity. The first two, INM2-F1 and INM2-F2 were launched from Cape Canaveral using McDonnell-Douglas Delta-II rockets in October 1990 and March 1991, respectively. INM2-F3 and

INM2-F4 were carried into space from the Kourou Space Centre in French Guiana using ESA Ariane rockets, in December 1991 and April 1992.

INM2 satellites incorporate a number of innovations in both vehicle construction and the communications package.

One new idea concerns satellite attitude stability and the way in which it 'sails' on what is euphemistically called the 'solar wind' in order to assist control. The two large solar arrays, which primarily provide all of the power to the electronic systems and subsystems on board the spacecraft, are fitted with extra 'flaps' to provide an increased surface area.

The on-board Attitude and Orbit Control System (AOCS), which is responsible for keeping the spacecraft stable and earth-pointing, gains information from various sensors placed on the body of the spacecraft. This AOCS subsystem then performs calculations for various parameters, resulting in rotational torque control of the solar arrays with respect to each other. The corresponding pressure of the solar wind on these surfaces controls the attitude stability of the spacecraft and maintains the antennae pointing towards the earth. Such control would normally be achieved through the use of several on-board thrusters, and the consequent saving of fuel normally needed to perform this function translates into achieving a longer operational life of the spacecraft.

The most obvious innovation in the communications system is that of the antenna farm. The curious cluster of cups forming each antenna, resembling a fly's eye and known as a cup-dipole array, is a method of reducing the size of the antennae. As an example of this, the large hexagonal cluster of 43 elements comprising the L-band transmit antenna is only 1.7 m in diameter.

A conventional parabolic antenna of the same specifications would need to be 30% larger. The antenna is the first direct radiating array on a satellite to use a shaped global beam. The effect is a subtle way to smooth out the power differentials which exist in a conventional beam as the effective received power reduces from the beam centre to its outer edges.

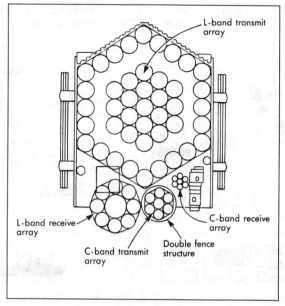

Fig. 5.7 Inmarsat 2 antennae. (Courtesy *Ocean Voice*)

Further innovations in the communications electronics enable the satellite to handle a wide variety of standards of mobile earth stations. Equally radical is the 180 W, L-band high power amplifier, with its four travelling-wave tube amplifiers (TWTA), which have been linearized to avoid interference from intermodulation noise and thus improve bandwidth usage. Another first is the use of surface acoustic wave (SAW) filters for channelization, permitting the transponder gain to be set independently for each standard of mobile earth station. The shore-to-ship transponder is divided equally into four sectors. One is devoted to the Inmarsat-A system, while another is used for the new Inmarsat-B service. Unidirectional high-speed data occupies a third, while the fourth is dedicated to the expanding Inmarsat-C service plus SAR and aeronautical services. INM2 satellites do, in fact, rotate on their solar panel axis once per day in order that the solar cells continue to face the sun whilst the antenna faces the earth. Power is stored in two nickel cadmium batteries to enable continuous operation whenever the satellite is eclipsed by the earth.

In common with all communications satellites INM2 is basically a signal repeater which is supported by control and environmental 'life support' systems. The satellite receives information from the land earth station in a directed beam of energy, at approximately 6 GHz in the C-band, converts the information and broadcasts it at, approximately, 1.6 GHz in the L-band, in a wide beam over one third of the earth's surface. The link from a mobile earth station follows the reciprocal path.

The ground segment

Preceding chapters in this book will have made the reader aware that all forms of telecommunications, including communications via satellite, must be rigidly controlled if total chaos is to be avoided. The same applies to the control of satellites in orbit. All satellites, irrespective of their function or nationality, must be rigidly maintained in their pre-arranged orbit. To satisfy both of these requirements an extensive ground control network has been established under the ultimate jurisdiction of Inmarsat.

In overall control of the whole network of fixed stations, mobile stations and satellites is the Inmarsat Network Control Centre (NCC), situated in London.

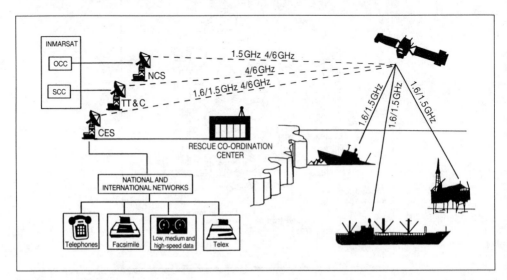

Fig. 5.8   Inmarsat-A ground operations

Three Satellite Control Centres (SCC) are responsible for the physical management of the four different types of satellites in use. Marisat and Intelsat V are controlled from the Intelsat and Comsat General SCC in Washington DC. Marec satellites are controlled from the ESA centre in Darmstadt, Germany, whilst the new generation INM2 satellites are controlled from the Inmarsat SCC in London.

Satellite tracking, telemetry and command (TT&C) is obviously of vital importance to ensure efficient satellite management. TT&C for Intelsat V, Marisat and Marecs space vehicles is initiated from their respective SCC. However, TT&C responsibilities for the new INM2 vehicles has been contracted out by Inmarsat to Telespazio at Fucino, Italy for the AORE and IOR satellites, to the China Satellite Launch and Tracking Control General company in Beijing for the POR craft and to COMSAT in Southbury and Santa Paula, in the USA, for the AORW and back-up services.

In this respect, Inmarsat was again at the forefront of new developments. Although these four TT&C stations are manned by site personnel during normal working hours, the equipment located at the stations is under the direct control of the SCC personnel at the London SCC. All reconfigurations of equipment on site at these stations, on a 24 hour basis, is nominally performed from the London SCC, together with the ability to repoint the large TT&C antennae remotely.

Each space craft is interrogated by a SCC to obtain the following data relating to its physical fitness:

- the operational status of vehicle subsystems;
- satellite orientation in space relative to the sun and the earth;
- a full diagnostic check on all electrical functions;
- the temperature of equipment and surfaces;
- quantity and availability of attitude control fuel.

At first sight this last item may seem to be superfluous. However, the useful life of the satellite is dependent upon its ability to maintain its attitude relative to the earth. The useful life can therefore be equated to the amount of fuel remaining for corrective manoeuvres to be carried out as required. The SCC is able, by the use of telemetry, to maintain a satellite attitude within $\pm 0.1°$.

Whereas the SCC is crucial to space vehicle management, the Network Co-ordination Station (NCS) is crucial to telecommunications services management. A NCS exists for each Inmarsat service and each ocean region. For the Inmarsat-A service, telecommunications control for the AORE&W is controlled by the Southbury NCS in the USA, the POR and the IOR by the Yamaguchi NCS in Japan. For the Inmarsat-C service, telecommunications control for the AORE&W is from the Goonhilly NCS in the UK, the POR by the Singapore NCS and for the IOR by the Thermopylae NCS in Greece.

A NCS continuously monitors traffic requests and the flow of telephone and telex traffic through the ocean region satellite for which it is responsible. This service is essential to maintain the correct operation between the mobile and the ground station.

The terrestrial end of the satellite communications link is made via a LES which forms the fixed end of the link.

Earth stations currently commissioned by Inmarsat are listed in figure 5.1.

A LES is a highly complex and very expensive telecommunications station to construct and operate. However, telecommunication services are able to generate large incomes and, as a consequence, new stations are becoming commissioned each year. There is no doubt that the list of LESs, serving the newer Inmarsat systems, will grow rapidly. The functions of a LES are to: establish communication channels in response to requests from terrestrial subscribers or mobile stations; to verify and file mobile station identities; to record traffic and process corresponding data and to identify distress

Fig. 5.9 Tangua Inmarsat-A CES. (Courtesy *Ocean Voice*)

priority calls from ships. LES services offered include: automatic calls for telephony and telex, manual operator services, directory enquiries, technical assistance, establish collect and credit-card calls, telegram services, store and forward services, group calls, data communications. A LES is also able to interface to value added services relating to health, navigation and other data.

A full detailed understanding of system operation and the interaction of each component can be gained from the next chapter in which descriptions of call establishment and procedure are detailed.

A land earth station

Land earth stations vary in size and complexity. The LES situated at Goonhilly Down in Cornwall, UK, is owned and operated by British Telecommunications (BT) the premier telecommunications provider for the United Kingdom. BT in the UK is an Inmarsat founder member and signatory with an investment share currently of 12.55480%. In addition to offering full Inmarsat-A and C LES facilities, Goonhilly is also the NCS for Inmarsat-C operations for the AORE&W ocean regions and offers the new Inmarsat-B and Inmarsat-M services.

It should be noted that the satellite earth station at Goonhilly Down provides a huge range of fixed link satellite services and, consequently, the Inmarsat mobile services are but a small part of the overall station. Inmarsat-A (AORW) services are currently based on antenna number 5 (GHY5), a 14 m Cassegrain dish. A second antenna provides the downlink for the AORE and a spare is provided in order to continue communications during maintenance.

The technical side of a typical maritime CES consists of three main features; the

Fig. 5.10 14 m Cassegrain Inmarsat dish link at Goonhilly. (Courtesy of BT)

antenna assembly, the radiocommunications electronics and the baseband signal processing system.

The parabolic antenna operates in both the L and C frequency bands to and from the satellite. It has previously been stated that the satellite up and down links with a CES occur at approximately 6 GHz in the C-band. However, a CES must be able to operate in the L-band to enable it to:

- monitor the CSC L-band channel and respond to requests for frequency allocations by the NCS;
- verify signal performance by loop testing between satellite and CES;
- receive the C-to-L band automatic frequency compensation (AFC) carrier.

In order to reduce the frequency errors seen by the SES, AFC is performed by a CES on the shore-to-ship (C-to-L frequency bands) communications signals. The primary reason for this action is to compensate for the frequency drift which occurs at the satellite and to maintain the accuracy of the received frequency at the SES.

The multiple effect of frequency error caused in the satellite transponder and the error caused by the Doppler frequency shift can result in frequency errors at the input to a SES approaching 55 kHz. The CES must include an AFC system whereby the error can be detected and corrected in order that the SES sees the correct L-band frequency from the satellite. To enable corrections to be made, common AFCs are transmitted by selected CESs for each of three sections of each ocean region—Northern, Equatorial and Southern. As an example, for the Northern AOR region the CES at Southbury in the USA transmits the AFC pilot carrier. After monitoring the carrier, each CES is able to correct the C-to-L band frequencies to within + 230 Hz relative to nominal and in the L-to-C direction to within + 600 Hz.

A typical Inmarsat-A CES antenna would be a Cassegrain structure with a dish of approximately 14 m diameter. Such an antenna would be able to meet the Inmarsat gain requirements, which are 50.5 dBi and 29.5 dBi respectively. The antenna is designed to withstand high wind speed, typically up to 60 mph in its operational attitude and 120 mph when stowed at 90°. The dish would be steerable ± 135° in azimuth and 0 to 90° in elevation. Tracking is either by automatic programme control or operator initiated. A tracking accuracy of 0.01° RMS and a repositioning velocity of $1° \text{s}^{-1}$ would be typical parameters for such a parabolic dish. The radio frequency and baseband processing hardware design varies greatly with station design and requirements.

Traffic accounting and billing

Traditionally, charges for traffic between ship and shore, and vice versa, have been split three ways; the ship station charge, the coast station charge and the landline charge, although the ship station charge has long since ceased to exist. This arrangement was not an even split of the cost involved in making a call. In addition, the charging method applied often involved three or more different administrations. For instance, the ship owner levied his charge, as did the telecommunications administration at the receiving end of the radio link, and finally there was the landline charge, possibly international, to complete the total. To this must now be added the cost of using Inmarsat satellite transponder time.

It should be noted that all forms of satellite telecommunication costings are based on the actual time, the 'connect time', that the call is in progress and not the 'holding time' or waiting period. The chargeable connect time for telex communications for example, is the period between receiving the answerback and the end of communications. All communications are therefore very accurately timed.

Traditionally, the laborious process of keeping radio traffic accounts fell on the shoulders of the radio officer, who took considerable pride in ensuring that the balance sheets (called abstracts) were neat, tidy and scrupulously accurate. With the demise of the radio officer on board GMDSS-equipped ships, it became evident that not only is electronics expertise being lost but also a working lifetime's experience of traffic routeing and accounting would go as well. Radio communications operation has now become a secondary task for maritime officers who obviously do not have the time or inclination to fill in lengthy radio abstracts. In order to alleviate the problem a number of traditional marine radio companies have devised computerized automated traffic logging systems. In some cases massive databases have been created to hold the world's complex traffic routes and charges. Databases are accessed, on a ubiquitous PC, via keyboard or touch screen commands. Radio abstracts are sent each month via a radio data link to the head office of the radio service accounting authority for processing.

Ultimately, the system will be fully integrated with the communications console and the whole process will become totally automatic.

Communications efficiency
Voice communication is, in fact, a very slow method of passing information when compared with other methods of communications, and is consequently inefficient. If telex or data transmission is used, the satellite transponder is active for only a short period and, consequently, the cost to transmit the same information could be a mere fraction of that quoted for voice communications.

As a simple example, assume that a single A4 page of information is to be sent over a communications link. A printed A4 page may contain 500 words of information. Each word consisting of an average of five characters.

Voice transmission
Using voice communications the page would take several minutes to send. Even longer when repeats are considered.

Telex transmission
In this mode each character consists of 1 start bit, 5 data bits and 1.5 stop bits (7.5 bits). Thus, the page of text would consist of 2500 characters corresponding to 18,750 bits.

The international telex network operational speed is 50 baud, which corresponds to a transmission speed of;

$$\frac{50 \text{ baud} \times 60 \text{ secs}}{7.5 \text{ bits}} = 400 \text{ characters/minute}$$

The page of text would now take 6 minutes 15 seconds to transmit.

This transmission time could be drastically reduced if the page was sent using facsimile or data transmission.

Facsimile transmssion
Standard Group 3 facsimile transmission would normally take 45 seconds for the A4 page. The normal set-up handshaking time for Group 3 machines is approximately 15 seconds—thus the total transmission time becomes 60 seconds.

Data transmission
Using data transmission each character plus control bits is a byte. The A4 page of data is now 2500 bytes. The message throughput for a 2.4 kbits s^{-1} modem is approximately 250 characters s^{-1}. Handshaking time for the modem is about 15 seconds giving a transmission time of 25 seconds.

It would appear therefore that the faster the data rate on transmission the shorter will be the transmission time. This is correct until the handshaking time is considered. High-speed modems tend to need longer periods for handshaking. As an example a 9.6 kbit s^{-1} modem needs approximately 30 seconds to perform handshaking. The message throughput is about 1000 characters per second, producing a total time of 32.5 seconds to handle the same A4 page of data. The chart in figure 5.11 compares transmission times using different transmission media.

In each ocean region there is more than one LES operating through the satellite. Each LES will be situated in a different country and will be controlled by a different telecommunications administration. Communications charges will therefore vary between LESs. It is possible for an SES operator to route his traffic through a CES which is cheaper than the others. In addition, the operator should make use of cheaper off-peak charges whenever possible.

A mobile making a call via the Inmarsat system involves several different elements. The LES through which the call is made calculates the cost of using all the elements

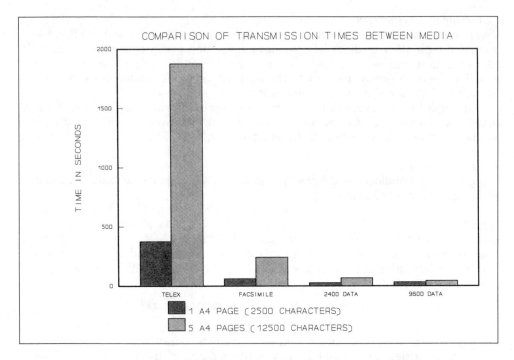

Fig. 5.11   Comparison of transmission times between media. (Courtesy Inmarsat)

involved and produces a total bill. The LES then invoices the call to an entity entitled the 'Accounting Authority' (AA) which has been previously nominated by the owners of the terminal and approved by national authorities. Accounting Authorities are identified by a code, termed the AAIC, consisting of four characters in two parts as follows:

(a) one or two letters indicating the country in which the AA is based;
(b) two digits to identify the specific accounting authority.

As an example NO01 indicates the first (01) AA in Norway (NO). The same codes are used in the terrestrial maritime mobile radio service.

In some countries direct billing to a 'Billing Entity'(BE) is permitted. The nomination of either a billing entity or an accounting authority must be agreed before a terminal can be commissioned into the Inmarsat system.

When a ship makes a call, its unique Inmarsat number IMN enables the LES computer to identify the AAIC or BE automatically. The AAIC consolidates all invoices from the different LESs used by a vessel over a given period. The AAIC then invoices the shipping company for the total amount out of which it pays the individual LES charges. Inmarsat invoices each LES directly for its use of the space segment.

To simplify billing and overcome difficulties which arise when a number of international administrations are involved, the process uses 'nominal' currencies. These are Gold Francs (GF) or the Special Drawing Right (SDR): 1 SDR = 3.061 GF. A CES calculates each vessel's traffic invoice using one of the nominal currencies and then invoices the AA concerned using a standard currency. The rate of billing applies at the date of the invoice and not at the date of the call which, of course, may be different. The conversion rate from the nominal currency to a standard currency depends upon the current exchange rate which is published by the International Monetary Fund (IMF).

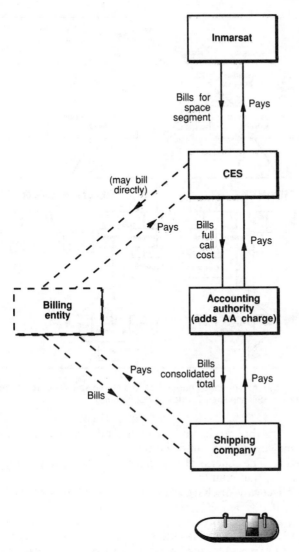

Fig. 5.12 The Inmarsat billing and settlement process

Coast earth station services

The telecommunication services offered by a LES vary greatly depending upon the complexity of the station selected. As an example of this, a typical LES along with the National Telecommunications Authority could offer a wide range of services from/to the mobile, located in their own ocean region to any international location. The Inmarsat-A services may include some of the following.

Telephone
- Distress, urgency, safety and medical assistance calls. Detailed in Chapter 1.
- Duplex telephone calls, either automatically or manually connected. Normal telephone calls to business or private subscribers.

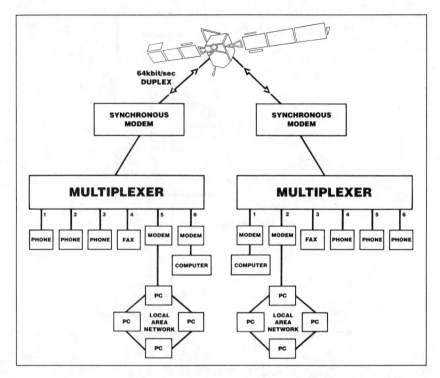

Fig. 5.13   Multiplexing six communications channels onto a single satellite connection using Inmarsat's duplex HSD service. (Courtesy *Ocean Voice*)

- Voice messaging. With this service a MES can send, retrieve, reply, redirect and broadcast spoken messages, even out of hours.
- National radio paging. The caller is able to page the called party using the National Radiopaging Service of that country.
- Cashless calling. Collect calls or charge cards may be used to pay for telephone calls.

Facsimile
MESs suitably equipped are able to send fax messages to a business fax number. Additionally, personal or private fax messages may be sent to a LES for forwarding by post.

Telex
- Standard telex point-to-point business connections, where the required telex receive number is not busy.
- Store and forward service. The message may be sent to the LES where the message will be stored, processed and forwarded over the telex network. Attempts will be made over the next 24 hour period to deliver the telex message.
- Multiple address telex messages. By using the store and forward service it is possible to send the same telex message to a hundred different destinations.
- Telex letter and greetings card. Greetings cards designed for special occasions.

Data and Information
- Databank. This supplies all the latest weather, financial and sports reports. Also essential maritime information such as navigation warnings and notices to mariners.

- Global network services (GNS). GNS offers access to international databases in over 80 countries. GNS gives access to a wide range of subjects including, meteorology, geoastrophys and oceanography. It is also possible to access electronic mail facilities and computing facilities.
- Electronic Mail. An electronic mail and information management service.
- Videotel. This is a Videotext service which is able to provide a vast amount of information including, meteorology, flight and ferry schedules, financial and business information.

Readers should check the documentation provided by LES operators and telecommunication authorities for details of these value added services. It should be noted that enhanced services like those above are not provided by Inmarsat but by an LES operator.

High speed data services
Inmarsat, along with a growing number of LESs, is now able to offer data communications at rates of 56 and 64 kbits s^{-1} compared with the much slower standard data rates of 2.4 kbits s^{-1} to 9.6 kbits s^{-1}. Two services are available, a ship-to-shore simplex demand assigned service and a full duplex 64 kbits s^{-1} service. The number of LESs able to offer these services is increasing although it should be noted that the service is dependent upon an adequate terrestrial link being available from the LES to the destination. This service appeals to high-volume data users such as research and exploration vessels.

Not all MESs can support the system but some modern equipment may be modified to do so. MESs which are able to do so are currently designated Inmarsat-A HSD or Inmarsat-A64/A56. Inmarsat-B terminals will offer a similar service in the future.

Multiplexing a number of communications channels onto a single satellite link, and thus greatly improving operational efficiency, becomes a real possibility by using duplex HSD. Figure 5.13 illustrates the principle.

6
The Inmarsat-A system

6.1 Introduction

The Inmarsat-A system has been the workhorse of maritime mobile communications since February 1982 when Inmarsat started to provide satellite communications for the marine user. Prior to that date a similar service was provided by the US COMSAT organization and was known as Marisat. The number of ship earth station fittings has massively increased since 1976. Demand for satellite channels has increased as a consequence, leading to the launch of newer, more powerful satellites with increased channel capacity. At the same time, electronic design technology has moved ahead to meet consumer demands for better, cheaper and more compact equipment. The net result is that an Inmarsat-A SES of 1994 is but a fraction of the size, weight and cost of those of 1982. The above decks equipment (ADE), including the radome of an SES, has shrunk in size and weight over the last decade to about a third of its original bulk. Owing to high volume sales and mass production techniques, the cost of a 1994 SES has also reduced to a fraction of what it was ten years previously, whilst the facilities it is able to offer have been greatly improved. A decade of development in the field of electronic design is equivalent to several decades in some other disciplines.

Outline system description

Inmarsat specifications provide for three classes of Inmarsat-A SESs as follows

- **Class 1:** Duplex telegraphy. (Telex).
  Shore-to-ship one-way telegraphy.
  Duplex telephony with and without compandors.
  Shore-to-ship one-way telephony with or without compandors.
- **Class 2:** Duplex telephony with or without compandors.
  Shore-to-ship one-way telephony with or without compandors.
  Shore-to-ship one-way telegraphy.
- **Class 3:** Duplex telegraphy.
  Shore-to-ship one-way telegraphy.

Speech compandors are used on a channel where voice communication is employed and not used when the same channel is to carry telex communications.

Technical specifications for equipment to be used via Inmarsat satellites are produced by Inmarsat for the guidance of equipment manufacturers. These specifications are extremely complex and are published in the document 'Technical Requirements for Inmarsat-A, Standard Ship Earth Stations' (or Inmarsat-B, C or M as appropriate) obtainable from Inmarsat headquarters in London.

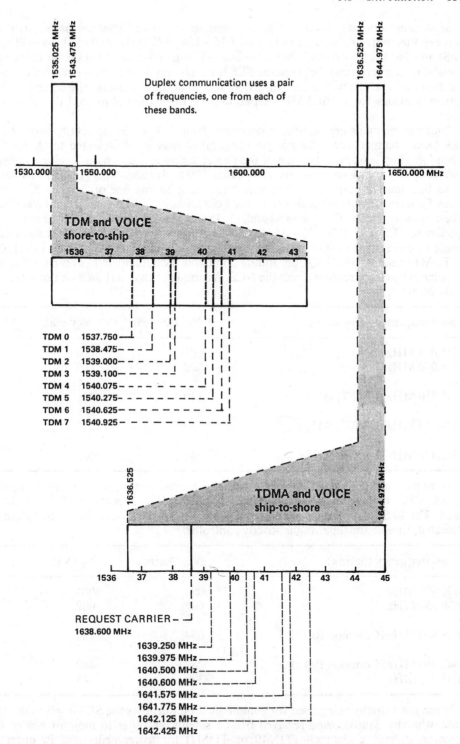

Fig. 6.1 Frequency bands for Inmarsat-A. (Courtesy ABB NERA AS)

SESs must operate, using assigned common frequency channel pairs anywhere between the limits of the two bands; 1535.025 – 1543.475 MHz (downlink—satellite-to-SES) and 1636.525 – 1644.975 MHz (uplink—SES-to-satellite). In order to reduce the possibility of interference between the SES transmission and reception, duplex communication uses one channel in each direction in each of the two bands of frequencies. The system provides for a 101.5 MHz separation between the uplink and downlink frequencies.

Inmarsat channels are numbered decimally from 001 to 339 or octally from 001 to 523. Octal channels carry the notation Q-octal or may be abbreviated to Q. An SES when initiating calls must use one of the two common request channels alternately for each call. The uplink channels are 124Q (1638.600 MHz) and 402Q (1642.950 MHz).

An SES must be capable of tuning automatically to any one of the 339 (523 octal, suffix Q) paired transmit/receive frequency channels in 25 kHz incremental steps anywhere between the limits of the two bands; 1535.025 MHz to 1543.475 MHz (downlink—satellite to SES) and 1636.525 – 1644.975 MHz (uplink). When in the idle state an SES monitors either of the common signalling channels (CSCs), TDM0 channel 110 (156Q) or TDM1 channel 139 (213Q), depending upon how the equipment has been configured, in order to listen for requests from the NCS. A sample of channels used on the downlink is shown next.

Radio frequency (downlink)	N_{10} Decimal	N_8 Octal
1535.025 MHz	001	001
1535.050 MHz	002	002
. . .	. . .	. . .
1537.750 MHz (CSC TDM0)	110	156
. . .	. . .	. . .
1538.475 MHz (CSC TDM1)	139	213
. . .	. . .	. . .
1543.475 MHz	339	523

To request a communications channel the SES must alternately use one of the two uplink, CSC frequencies 1638.600 MHz or 1642.950 MHz, allocated within the uplink band. The SES transmit RF channel is always 101.5 MHz above the receive channel allocated, thus channel pairing is strictly controlled.

Radio frequency (uplink)	N_{10} Decimal	N_8 Octal
1636.525 MHz	001	001
1636.550 MHz	002	002
. . .	. . .	. . .
1638.600 MHz (Common RQ ch.)	084	124
. . .	. . .	. . .
1642.950 MHz (Common RQ ch.)	258	402
1644.975 MHz	339	523

When not actually being used for communication purposes the SES is left in the idle state, whereby data is being received from a satellite in order to monitor one of the common signalling channels (TDM0 or TDM1) for assignments and in order to maintain satellite lock.

There are two basic types of signalling in the Inmarsat-A system. Out-of-band

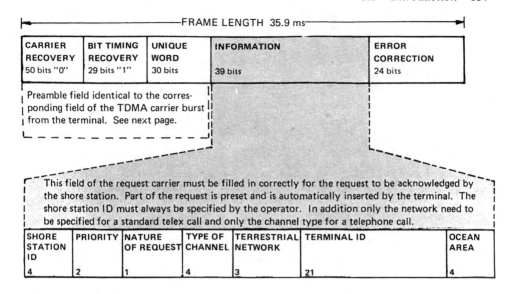

Fig. 6.2   Request burst format. Ship-to-shore

signalling, which is used to set up and control the various types of communication channels via the satellite and in-band signalling, which is used for all supervisory and selection signalling once the satellite channel has been assigned.

Out-of-band signalling to the SES is contained in the 'signalling channels' of the TDM carriers transmitted by the NCS and the CES. There is at least one TDM carrier continuously transmitted by each CES. Additionally, there is also a common signalling carrier (CSC) continuously transmitted on the selected idle listening frequency by the NCS. The SES remains tuned to the CSC at all times even when the SES is in the idle state. The carrier is modulated with a rectangular BPSK data waveform at 1.2 kbits s^{-1}.

When requesting communication channels, the SES must transmit a request carrier burst on one of the two assigned CSCs. Modulation is coherent BPSK with a data rate of 4.8 kbits s^{-1}. NRZ coding format is used to reduce bandwidth. The EIRP of a single carrier from ship to satellite is 36 dBW and remains constant.

To ensure that there is no cross channel interference when communicating by telex or data, an 'Interchannel Gap' is provided of 41.4 ms. The time delay appears to be excessively large when compared with the 37.7 ms information frame when using telex. Obviously, if this delay could be reduced a greater number of channels could be made available within a data frame. The interchannel gap delay of 41.4 ms has been calculated as a result of differences in signalling distances between ships, the satellite and the CES. The difference in path distances from the best case, where the SES and the CES have a combined path length of approximately 72,000 km, to the worst case where the SES, in high latitudes, and a given CES have a path length of approximately 78,000 km, produces a difference in signal delay of about 50 ms. This figure is a very broad indication of the signal delay over 6000 km for a radiowave travelling at 300 x 10^6 m s^{-1}. In practice, the maximum delay likely to occur has been accurately calculated to be 41.4 ms. If the interchannel gap was reduced below this figure, ships communicating in the worst case scenario may suffer synchronization problems causing channel time slots in

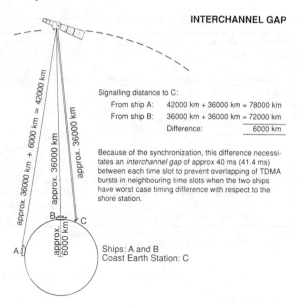

INTERCHANNEL GAP

Signalling distance to C:

From ship A: 42000 km + 36000 km = 78000 km
From ship B: 36000 km + 36000 km = 72000 km
                Difference: 6000 km

Because of the synchronization, this difference necessitates an *interchannel gap* of approx 40 ms (41.4 ms) between each time slot to prevent overlapping of TDMA bursts in neighbouring time slots when the two ships have worst case timing difference with respect to the shore station.

Ships: A and B
Coast Earth Station: C

Fig. 6.3   Explanation of the interchannel gap. (Courtesy ABB NERA AS)

the frame to overlap with neighbouring time slots. Clearly this would result in cross channel interference. Figure 6.3 illustrates the worst case scenario where two ships are communicating with a CES over very different distances.

Call initializing and clearing procedures

Call initialization and clearing for telephone working
Telephone calls and voice-band data calls (medium and low speed data and fax) are transmitted over the network using FM channels. It should be noted that the procedure detailed below is fully automatic and only requires the SES operator to input the dialling sequence of numbers to make a call. The description which follows assumes that the SES has been tuned to monitor the CSC TDM0 channel 156Q.

The process is started by the operator who selects a CES and initiates a request data burst using either of the two request channels. The SES will automatically alternate request bursts between the two allocated request channels, channel 124Q or 402Q, both of which are continuously monitored by all CESs and the NCS in the ocean region. The NCS monitors these channels in case there is no response from a CES to a priority 3 (distress) request in which case a co-located NCS will initiate a call.

An operator dials the wanted CES, in this case 0411 Eik, Norway, and thus initiates a 'request burst signal' containing the type of communication required, the CES identity, the priority of the communications (priority 3 or routine), the type of channel if for telephony (companded) and the identity number of the SES. The addressed CES reads the request and sends a 'request for the telephony channel assignment' message, on channel 244Q, to the NCS. The NCS will assign a channel and announce the assignment to both the CES and the SES over the CSC 156Q. On receipt of the assignment from the NCS the CES will tune one of its modems to the assigned channel and switches on both an FM carrier and a single frequency (SF) tone (2600 Hz). The SES tunes to the assigned channel and waits for a signal from the CES. On receipt of the carrier and the SF tone from the CES the SES will turn on its carrier and modulate with an SF tone. When the

TELEPHONE SHIP – SHORE

The example below shows a telephone connection via EIK coast earth station, Norway.

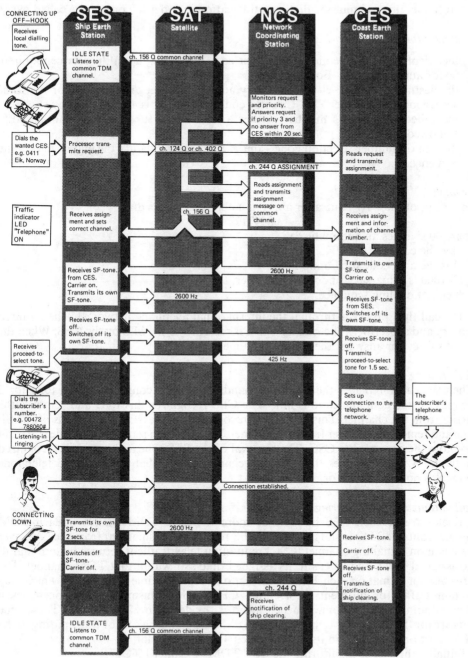

Fig. 6.4 Inmarsat-A telephone channel call set-up and clearing (ship originated). (Courtesy ABB NERA AS)

CES receives the SF tone from the SES it removes the SF tone from its own transmission. The SES recognizes the removal of the SF tone from the CES frequency and will turn off its own SF tone. This completes the 'handshaking' process causing the CES to send a 'Proceed to select' (PTS) signal to the SES. This tone of 425 Hz lasting for 1.5 s informs the SES operator to input the dialling information in order to contact the called party.

Dialling information comprises

- the telephone service code—a two-digit code depending upon the service required. 00 for automatic calls. Followed by,
- the destination code—either a country access code (see Appendix 4) or maritime access code for another SES. Country code for the UK is 44, followed by,
- the called subscriber's number—either a land based subscriber or another SES. Followed by,
- the end of number selection code—the # character to signify the end of the calling sequence.

Example
On receipt of the PTS Tone after approximately 12 seconds dial
0044717281000
where 00 signifies an automatic call,
     44 is the code for the UK,
     71 is the area code for London,
     7281000 is Inmarsat London HQ number,
     # end of dialling sequence.

The call then proceeds through the international telephone network to the required country and subscriber. The ringing signal is returned to the SES by the CES. When the called subscriber answers, the chargeable time starts.

Call clearing
When the caller replaces the telephone handset the SES connects the SF tone for 2 seconds on the communications link. The CES recognizes this tone and removes the carrier from the channel. This is, in turn, recognized by the SES which turns off its SF tone and removes its carrier from the channel. The SES then returns to the idle state and monitors the appropriate CSC. The removal of the SES tone and carrier is recognized by the CES which notifies the NCS, on channel 244Q, that the occupied telephone channel is now available for re-assignment.

Call initialization and clearing for telex working
For telex traffic each CES has one or more exclusive TDM channels. Each TDM channel contains 22 time slots. When an SES receives an assignment from the CES over the common signalling channel (CSC) requesting telex traffic it retunes its terminal to the assigned TDM and listens to its assigned time slot within that TDM channel. The transmission format on each TDM is split into 22 separate time slots each of which may contain traffic for different SESs. Each CES TDM transmit channel possesses a corresponding receive channel, which again is divided into 22 time slots. These time slots are for the CES receive channel. The SES, when handshaking or transmitting to the CES, must only transmit in its own assigned TDMA time slot.

Initially, the SES is monitoring the CSC TDM0 channel 156Q.

When the operator makes a request to a CES, in this case 041 Eik, Norway, the processor transmits a 'request burst' on channel 124Q or 402Q, to the addressed CES.

22

TELEX SHIP – SHORE

The example below shows a telex connection via EIK coast earth station, Norway.

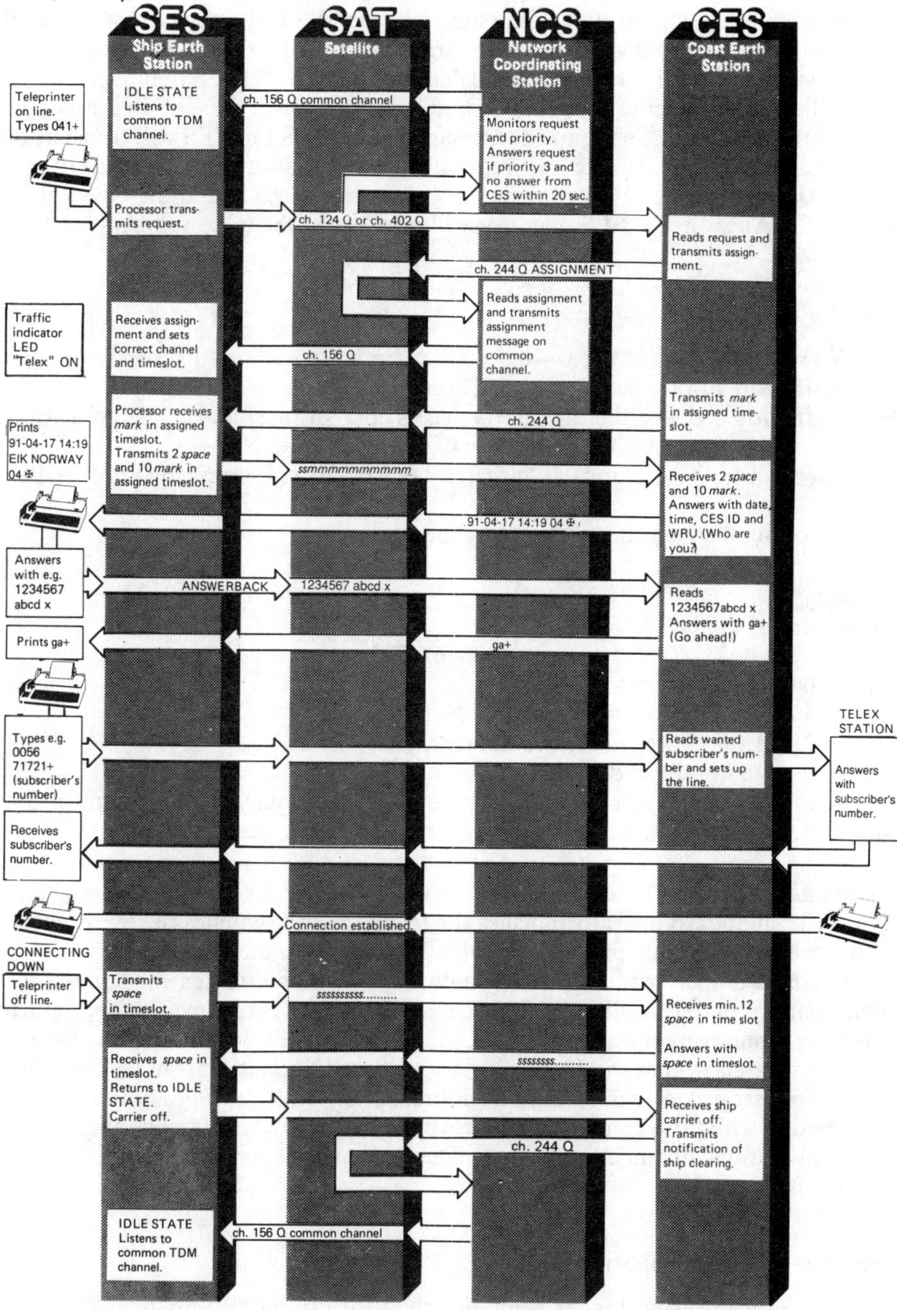

SES Ship Earth Station   **SAT** Satellite   **NCS** Network Coordinating Station   **CES** Coast Earth Station

Teleprinter on line. Types 041+	IDLE STATE Listens to common TDM channel.

ch. 156 Q common channel

Monitors request and priority. Answers request if priority 3 and no answer from CES within 20 sec.

Processor transmits request.

ch. 124 Q or ch. 402 Q

Reads request and transmits assignment.

ch. 244 Q ASSIGNMENT

Reads assignment and transmits assignment message on common channel.

Traffic indicator LED "Telex" ON

Receives assignment and sets correct channel and timeslot.

ch. 156 Q

Transmits *mark* in assigned timeslot.

Prints 91-04-17 14:19 EIK NORWAY 04 ✠

Processor receives *mark* in assigned timeslot. Transmits 2 *space* and 10 *mark* in assigned timeslot.

ch. 244 Q

ssmmmmmmmmmmm

Receives 2 *space* and 10 *mark*. Answers with date, time, CES ID and WRU.(Who are you?)

91-04-17 14:19 04 ✠

Answers with e.g. 1234567 abcd x

ANSWERBACK   1234567 abcd x

Reads 1234567abcd x Answers with ga+ (Go ahead!)

Prints ga+

ga+

Types e.g. 0056 71721+ (subscriber's number)

Reads wanted subscriber's number and sets up the line.

TELEX STATION

Answers with subscriber's number.

Receives subscriber's number.

Connection established

CONNECTING DOWN

Teleprinter off line.

Transmits *space* in timeslot.

sssssssssss.........

Receives min. 12 *space* in time slot

Receives *space* in timeslot. Returns to IDLE STATE. Carrier off.

sssssss.........

Answers with *space* in timeslot.

Receives ship carrier off. Transmits notification of ship clearing.

ch. 244 Q

IDLE STATE Listens to common TDM channel.

ch. 156 Q common channel

Fig. 6.5 Inmarsat-A telex channel call set-up and clearing (ship originated). (Courtesy ABB NERA AS)

The burst signal contains the type of communication required (telex), the priority and the identity number of the SES. The CES will assign one of the time slots on his/her TDM to the SES. This assignment is passed to the NCS on channel 244Q, for retransmission to the SES on the CSC in order to set the correct channel and timeslot for communication. On receipt of this assignment the SES will retune to the assigned TDM and, while listening to that time slot and transmitting into a corresponding time slot, will carry out satellite call transition (handshaking) with the CES. Satellite call transition is the exchanging of a series of '2 marks' and '10 spaces' with the CES in the assigned time slots. After passing the assignment message to the NCS the CES will transmit the date, UTC time and CES identity and request the SES answerback by transmitting WRU (Who Are You?). The SES will respond with its answerback and the CES will transmit GA + (Go Ahead). The SES operator will now key in the telex numbers for connection to be made.

Dialling information comprises

- Telex service code—a two-digit code depending upon the service required. 00 signifies an automatic connection. Followed by,
- destination code—either a country access code or maritime access code for another SES. Telex country code for Norway is 56. Followed by,
- called subscriber's number—either a land-based subscriber or another SES. Followed by,
- end of number selection—the ' + ' character.

Example
On receipt of the GA + signal dial

005671721 +

where 00 signifies an automatic call,

56 is the telex country code for Norway,

71721 the shore subscriber number,

\+ signifies the end of the sequence.

The call is then connected and the chargeable time starts from receipt of the called subscriber's answerback.

Telex call clearing
When the shipboard operator finishes the telex call and switches the telex machine 'Off Line', the SES will exchange a series of '12 spaces' with the CES.

The CES will then send to the SES confirmation of satellite clearing. On receipt of this confirmation, the SES will turn off its carrier in the assigned time slot and return to the idle state monitoring the appropriate CSC. The CES will then inform the NCS that the SES has cleared so that the NCS can remove the SES identity from its 'Ship Busy' list.

The two sequences describing call initialization and clearing should be studied in conjunction with the SES manufacturer's operator manual. Reference may also be made to the operations documentation issued by National Communications Authorities for mobile users.

Inmarsat-A SES equipment

Whilst equipment is no longer being designed for use on the Inmarsat-A system there are tens of thousands of Inmarsat-A SESs in use in the world and, consequently, the system will be in use for many years to come.

The council of Inmarsat has adopted the following policy in relation to Inmarsat-A services:

- the services will be supported until at least the years 2005/2006;
- Inmarsat will continue to support new commissionings of Inmarsat-A SESs for a minimum of three years after the introduction of the Inmarsat-B and Inmarsat-M systems;
- Inmarsat will continue to support the re-commissioning of Inmarsat-A SESs after the introduction of the Inmarsat-B and M systems.

Indeed, in July 1993 the Inmarsat-A Network Control Stations were extensively redesigned in order to support a major change in system access and improve the allocation of communication channels. The original Inmarsat-A NCSs were commissioned some 15 years earlier and, as a consequence, their electronic processing capability has become slow and limited when compared with modern systems. In addition to improving electronic processing the new NCSs are capable of channel interleaving/ inserting in order to improve greatly access time and channel efficiency.

Originally, all Inmarsat-A SESs communicated with the appropriate ocean region NCS on the common signalling channel (CSC) TDM0 in order to request a communications channel. A situation which caused queues to be formed during busy periods resulting in unnecessary waiting periods. From July 1993 there are two common signalling channels available for use by SESs in an ocean region, known as TDM0 and TDM1. All SESs commissioned by that date are allocated alternately one of the two CSCs. This, in effect, means that the access time for an SES may be reduced by half under those conditions where two ships require access simultaneously, assuming of course that the two SESs are each allocated one of the two different CSCs: TDM0 and TDM1.

In order to ensure that an access clash is unlikely, approximately half of the world's Inmarsat-A SESs were re-tuned to the CSC TDM1 whilst the other half remained tuned to TDM0. The criteria used to determine which SES was re-tuned centred on the unique Inmarsat mobile number (IMN) allocated to an SES. Those SESs which possess an even digit as the fourth digit of the INM number remain tuned to TDM0. Those with an odd digit for the fourth digit retuned to TDM1. The situation continues with new commissionings.

Because Inmarsat-A SESs are monitoring different CSCs, distress alerting and message information is broadcast to SESs on both CSCs. Whilst the modifications to the system are fully transparent to the user, some SESs will revert to the original CSC after being switched off. Re-tuning is fairly straightforward at switch on. Refer to the manufacturer's manual.

Another Inmarsat-A enhancement is to customize the uncompanded channel 02 for data and facsimile calls. Customizing this channel required significant alignments to be made at each CES and SES. The circuit alignments, which must not be carried out by unauthorized personnel, involve the limiting of the audio signal level to the frequency modulator in order to limit frequency deviation. For further details refer to Frequency Modulation in Chapter 14.

There are currently 13 manufacturers of Inmarsat-A fixed equipment some of whom also manufacture the Inmarsat-B, C and Inmarsat-M systems. Because of the extensive design, development and type approval costs, Inmarsat-A equipment manufacturers tend to be large International companies. Such companies are more able to provide the full world-wide maintenance support which an SES may need in ports throughout the world.

Appendix 3 lists the producers of Inmarsat terminals.

Fig. 6.6   Japan Radio Company Inmarsat-A radome fitted on a modern vessel. (Courtesy JRC/Raytheon)

Equipment type-approval and commissioning

Inmarsat, quite rightly, operates an extensive type-approval service through which a proposed Inmarsat terminal must pass before being allowed access to the space segment. The document 'Technical Requirements For Inmarsat Ship Earth Stations' sets out rigid technical standards for the proposed equipment. Once the equipment has passed through the design and development stage, a pre-production unit is sent for standards testing. If all is well and the equipment passes, the manufacturer goes into production and hopefully sells sufficient units to cover the extensive development costs. Each MES sold is professionally installed, commissioned by Inmarsat and issued with a unique Inmarsat mobile number (IMN) previously known as the terminal ID. A number of Inmarsat-A terminals possess a second IMN which enables users to receive and set-up calls to two separate subscribers. However, it is not possible for a Dual-IMN SES to carry more than one link call simultaneously, as only one communications channel is available for use.

A multi-user Inmarsat-A terminal possesses the ability to operate two simultaneous link calls using two channels. A small number of Inmarsat-A approved SESs offer this facility.

SES system description

Essentially, Inmarsat-A SES equipment has two parts, the above decks equipment (ADE) and the below decks equipment (BDE). The ADE is easily recognizable by the large opaque radome which provides environmental protection for the unit.

In general an ADE consists of the following.

• **The Parabolic Antenna**—On older equipment this may be in excess of 1.2 m in diameter, whereas on newer equipment it is likely to have reduced to approximately 0.8 m. Ever since the introduction of Inmarsat-A SES equipment the practice has been to reduce the size of the antenna and, consequently, the size and weight of the ADE. As a large proportion of the receiver signal gain and transmitter EIRP is produced by the antenna, the area of the dish can only be reduced if the transmitting power from the satellite transponder is increased and/or the gain of the receive pre-amplifiers can be increased without an appreciable increase in noise.

• **The Stable Platform**—This is the antenna support assembly which must remain perfectly stable when the ship is pitching or rolling in extreme weather conditions. It is essential that the stable platform holds the dish antenna in its correct AZ/EL angular positions despite movements of the ship. The platform usually consists of a large solid bed mounted in such a way that four gyro compasses are able to sense movement and correct any errors which are detected, holding the platform level. It is, in practice, a form of electronic gimbals.

• **The Tracking System**—The dish antenna is controlled in AZ/EL by stepping motors which, in turn, are electronically controlled in a simple feedback system. This electro-mechanical arrangement enables the dish to maintain a lock on a satellite despite navigation course changes. As the ship changes course both AZ/EL control corrections must be made.

• **The Computer**—The ADE processor controls all ADE functions, which include satellite tracking and electronic control.

• **The RF Electronics**—These contain the transmitter high power amplifier (HPA) and the receiver RF 'front end' low-noise pre-amplifier stages (LNA), plus all the critical bandpass signal filter stages.

• **The Multiplexer**—In modern equipment it is common practice to reduce the number of cables between ADE and BDE. This is achieved by multiplexing up/down signals or commands between ADE and BDE onto the one coaxial feeder.

The BDE houses the following units.

• **The System Computer**—This processor controls the whole system operation although some of its functions may be distributed to second-level dedicated processors.

• **Modulator/Demodulator**—This forms the nucleus of any communications system. In this case it modulates baseband signals onto the intermediate frequency (IF) carrier and extracts the baseband signals from the IF carrier of the received signal.

• **The Synthesizer**—This produces the highly stable frequencies required for modulation/demodulation and signal switching.

• **Audio Processor**—This performs baseband processing of the audio signals from the microphone on transmit and from the demodulator output on receive.

• **Data Processor**—This supervises data processing of both transmit and receive telex signals. It monitors out-of-band signals.

• **Interfacing**—Interfacing performs all the tasks of signal conditioning between peripherals and signal processing.

• **Operator Display and Control**—System functional control may be achieved by using a dedicated control panel as was often the case in older units or, in modern equipment, by using a standard typewriter keyboard on an IBM PC clone. Increasingly, manufacturers are using versatile off-the-shelf computer terminals for control instead of designing a dedicated control panel.

The preceding section broadly outlines the systems incorporated in an Inmarsat-A SES terminal. As each manufacturer inevitably produces a different design to those of

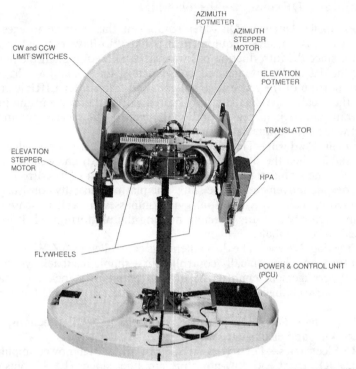

Fig. 6.7   An ADE stabilized antenna assembly minus the radome. (Courtesy ABB NERA AS)

the competitors and, although SESs tend to look very different, they do in fact perform the same functions.

SES ADE antenna specifications and siting

The communications antenna of an Inmarsat-A SES is of necessity relatively large and heavy. Over the past decade the ADE—which comprises the mechanical antenna assembly, the antenna control electronics and gyroscopes, and the microwave electronic package and the dish antenna—has reduced considerably in both physical size and weight. This reduction, brought about by greater EIRP from satellite transponders coupled with the use of high-gain/low-noise GaS-FET technology at the front end of the receiver has made the fitting of Inmarsat-A SES terminals on small craft a reality. Some of the very early ADEs weighed several hundred kg which, when mounted 6 m above the top deck of a vessel led to severe stability problems. Currently, an ADE for a SES is likely to weigh less than 100 kg and include a parabolic antenna of approximately 0.8 m in diameter.

The technical principles of parabolic antennae are detailed in the companion volume, *Satellite Communications: Principles and Applications.* Inmarsat does not specify the physical size of an antenna or ADE it but does specify the signal parameters which must be achieved, amongst which are the following.

● **Steerability and Tracking:** the antenna beam must be capable of being steered in the direction of any geostationary satellite whose orbital inclination does not exceed 3° and whose longitudinal excursions do not exceed ± 0.5°. Means must be provided to point the antenna beam automatically towards the satellite with sufficient accuracy to ensure that the transmit and receive signal levels are satisfied continuously under operational conditions.

Careful consideration should be given to the siting of an Inmarsat-A radome. Essentially, the focal point of the parabolic antenna must be pointing directly at the geostationary satellite being tracked without any interruption of the microwave beam, which may be caused by obstructions on the ship. Inmarsat specify that there should be no obstacle which is likely to downgrade the performance of the equipment in any angle of azimuth down to an elevation of $-5°$. This is not easy to achieve. Figure 6.8 illustrates a theoretical antenna installation satisfying these guidelines but with the disadvantage that it is very high above the vessel's deck and would be impossible to install. This antenna would certainly be adversely affected by vibration and it would be difficult to gain access for maintenance purposes.

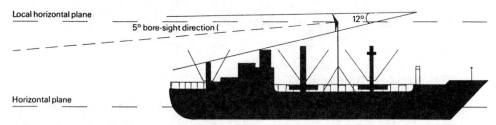

Fig. 6.8   Ship earth station design and installation guidelines. (Courtesy Inmarsat)

If structures do interrupt the communications beam, blind sectors will be caused leading to degraded communications over some arc of azimuth travel. However, it should be remembered that the communications RF beam of energy possesses a width ($12°$ angle cone) and consequently objects within 10 m of the radome, which cause a shadow sector greater than $6°$ are not likely to degrade the performance of the equipment significantly. It is preferable that there should be no objects within 3 m of the antenna. Beyond this distance, small objects measuring less than 15 cm can be ignored.

If, as is often the case, it is impossible to find a mounting position free from all obstructions, any blind sectors identified should be recorded. It may be possible for the operator, when in an area served by two satellites, to select the satellite whose azimuth and elevation angles with respect to the ship are outside the blind sector.

Electromagnetic RF signals at high radiation levels are known to be hazardous to health. It is inadvisable to permit human beings to stand close to the radome of an SES when the system is communicating with a satellite at a low elevation angle. It is not possible to quote a safe distance because international regulations concerning this subject vary. Inmarsat recommends that the radiation levels in the vicinity of the antenna should be measured. The distances from the antenna which result in radiation levels of 100 W m^{-1}, 25 W m^{-1} and 10 W m^{-1} should all be recorded. Radiation plan diagrams may be produced and located near the antenna as a warning, or distances from the antenna may be physically labelled.

Satellite location and tracking

An Inmarsat-A, B or M mobile antenna must be capable of locating and continuously tracking the geostationary satellite selected for communications. Inmarsat C uses an omnidirectional antenna. Locating and tracking may be done automatically as in the case of a SES installation, or manually only, as with a portable system.

Direction (azimuth angle)

The azimuth angle is the angle between north and the horizontal satellite direction as seen from the MES.

The actual azimuth angle for the various satellites due to the Saturn terminal can be found on the map on page 149.

Example:
259° azimuth

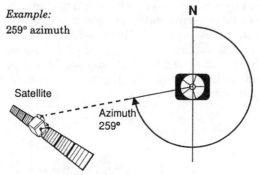

Height (elevation angle)

The elevation angle is the satellite height above the horizon as seen from the MES.

The actual elevation angle for the various satellites due to the Saturn terminal can be found on the map on page 150.

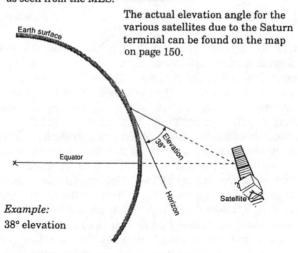

Example:
38° elevation

Fig. 6.9   AZ/EL angle explanations. (Courtesy ABB NERA AS)

It is common practice to believe that the satellite is 'fixed' and that once the link has been established it will remain so as long as the mobile does not move. It should be remembered that the satellite is not stationary, it is, in fact, travelling at some 3.073 km s^{-1} and is geostationary when viewed from the surface of the earth. A satellite is under the influence of a number of variable astrophysical parameters which cause it to move around its station by up to several degrees. A mobile tracking system must therefore counteract this by repositioning the MES antenna at regular intervals. The carrier signal is monitored continuously and, if a reduction in its amplitude is detected, a close programmed search is initiated until the carrier strength is again at maximum. No loss of signal occurs during this process, which is automatically initiated.

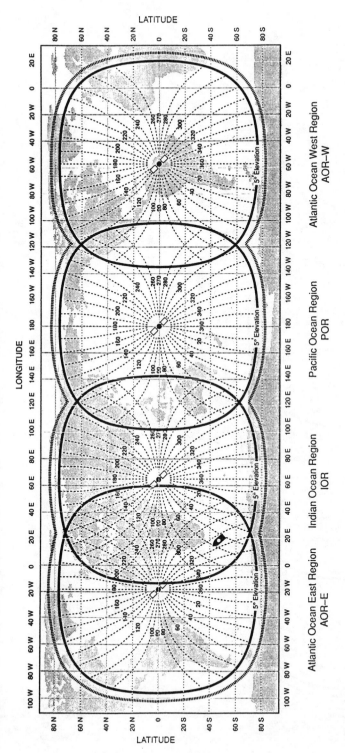

Example:

Azimuth angle for the plotted position

315° for the AOR-E satellite
55° for the IOR satellite

Be careful not to read the wrong angle in areas
where two satellites overlap.

Fig. 6.10 (a) Azimuth angle map. (Courtesy ABB NERA AS)

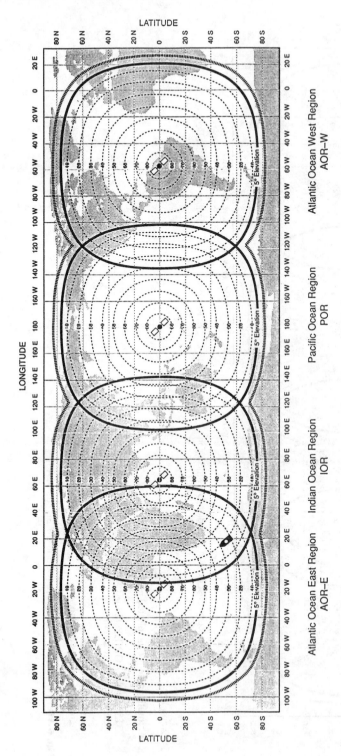

Example:

Elevation angle for the plotted position

24° for the AOR-E satellite
17° for the IOR satellite

Be careful not to read the wrong angle in areas
where two satellites overlap.

Fig. 6.10 (b) Elevation angle map. (Courtesy ABB NERA AS)

Obviously, the greatest tracking problem will arise when the mobile station is moving at speed with respect to the satellite. An Inmarsat-A SES antena may be moved through any angle in azimuth (AZ) or elevation (EL) as the vessel moves along its course.

It is essential therefore that electronic control of the antenna is provided. In practice, a simple electronic feedback control system is used. Control may be achieved either by manual or automatic methods.

● **Manual commands.** When the operator has selected manual control, elevation is commanded by Up and Down keys, whereas azimuth positioning is controlled by Clockwise (CW) and Counter Clockwise (CCW) keys. This command would be used when the relative positions of both the vessel and the satellite are known. AZ/EL angles can be derived and input to the equipment, by using the two AZ/EL charts provided by ABB NERA. See Figures 6.9 and 6.10.

Once the antenna starts to detect a satellite signal the operator display indicates signal strength. Fine positioning can now be achieved by moving the antenna in AZ/EL in 1/6th degree steps until maximum field strength is achieved.

● **Automatic control.** Once satellite lock has been achieved the system will automatically monitor signal strength and apply AZ/EL corrections as required in order to maintain this lock as the vessel changes course.

● **Automatic search.** An automatic search routine commences 1.5 minutes after switching on the equipment or it may be initiated by the operator. The EL motor is caused to search between 5° and 85°, whereas the AZ motor is stepped through 10° segments. If the assigned common signalling channel signal is identified during this search the step tracking system takes over to swing the antenna above/below and to each side of the signal location searching for maximum signal strength. Figures 6.11 to 6.13 show the ordinary step track and the fast step tracking sequences where the antenna is oscillated either side and above/below the beam until maximum signal strength is achieved.

● **Gyroscopic control.** Lock is maintained irrespective of changes to the vessel's course by sensing signal changes in the ship's gyro repeater circuitry. Satellite signal strength is monitored and, if necessary, the EL/AZ stepper motors are commanded to search for maximum signal strength.

● **Antenna rewind.** The antenna in the ADE is pivoted on a central mast and is coupled by various control and signal cables to the stationary stable platform. If the antenna was

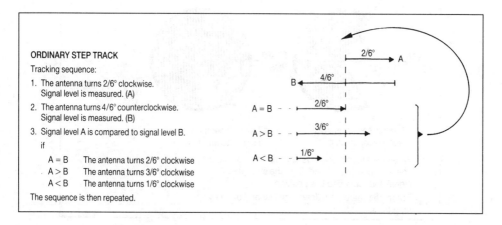

Fig. 6.11   Antenna ordinary step-track sequence. (Courtesy ABB NERA AS)

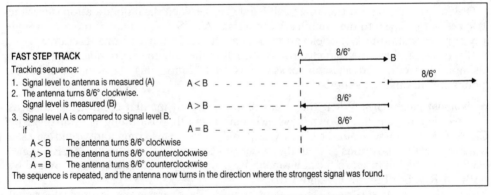

Fig. 6.12   Antenna fast step-track sequence. (Courtesy ABB NERA AS)

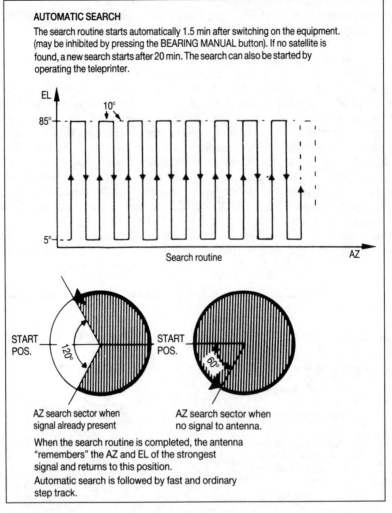

Fig. 6.13   Automatic search. (Courtesy ABB NERA AS)

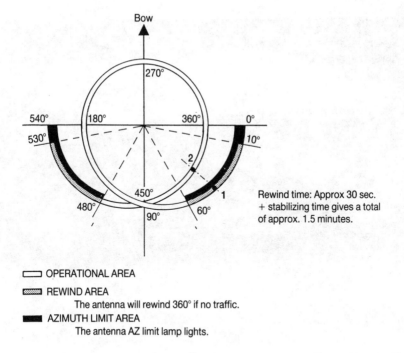

Fig. 6.14   Antenna azimuth limit. (Courtesy ABB NERA AS)

permitted to rotate continuously in the same direction, the feeder cables would eventually become so tightly wrapped around the central support that they would either prevent the antenna from moving or they would fracture. To prevent this happening a sequence known as antenna rewind is necessary. Figure 6.14 illustrates the process.

In fact, the antenna is permitted to rotate through 540° in azimuth, although the antenna would normally operate within the segment 60° to 480°. If the antenna moves beyond its operational area into one of the designated rewind areas, 10° to 60° or 480° to 530°, and if no traffic is in progress, the antenna will automatically rewind through 360°. In figure 6.14 the antenna would automatically move from position 1 to position 2, the same AZ/EL angle with the satellite, to maintain lock.

If the SES is processing a call when the antenna runs into the rewind area, no rewind will occur unless it continues into the AZ limit regions. If this happens rewind will take place and the call will be lost. The AZ warning indicator on the operator display will light to indicate that antenna rewinding is in progress.

The necessity for antenna rewind is absolute. Using current technology it is not yet possible to have any other method of connection between antennae than cable feeds. Receive signal levels are extremely small and, consequently, vulnerable to high loss if a system of slip rings or some form of local transmission system was used. The problem of rewind does not apply to transportable Inmarsat-A equipment where the whole unit is manually rotated to point at a satellite. Clearly, it would not be possible to fix the antenna of a SES and swing the ship in order to achieve lock! However, it should be noted that the use of an Inmarsat-A transportable land MES in a maritime environment is not permitted as it is not type approved for a marine environment and is incapable of generating priority 3 distress alerts

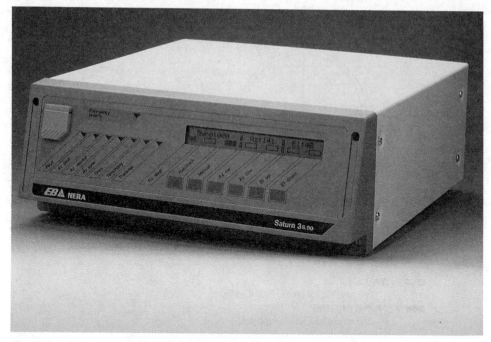

Fig. 6.15   The Saturn 3S.90 SES. (Courtesy ABB NERA AS)

ABB NERA AS—Saturn 3 SES

ABB NERA AS Satcom Marine, based in Billingstadsletta, Norway, is a major manu-facturer of all Inmarsat standards of equipment. The company was, in fact, the first European company to design and manufacture SESs. Currently, nearly 3000 Saturn terminals are in service on ships, on land and on offshore oil platforms.

ABB NERA also manufacture and supply CESs. The CESs at Eik, Thermopylae, Jeddah, Beijing and Perth were all supplied by this large company. Aeronautical GESs in the UK, Singapore, Norway, Australia, France, Germany and the USA are currently being supplied by ABB NERA. The Saturn 3S.90 which is fully compatible with the GMDSS requirements.

When in the idle state the system continuously receives the assigned CSC, in this case TDM0 on 1537.750 MHz. The duplex filter diverts the received carrier from the output of the transmitter high-power amplifier (HPA) and applies it to the down converter via the low noise amplifier (LNA). The ability to amplify the minute signal from the satellite without increasing noise level is particularly important in all satellite receiving systems. The required signal from the dish antenna is likely to be -128 dBm (0.1 μV) which must be amplified before processing can take place. All active devices produce noise. If noise is added to the signal at this stage it will simply be amplified along with the signal throughout the receiver. The LNA contains active devices made from gallium arsenide (GaS FET) which have been proved to be very efficient low-noise devices.

The down converter is a signal mixer which mixes the amplified TDM signal with a locally generated frequency to produce the nominal intermediate frequency (IF) of 210 MHz. A locally generated frequency 1330 MHz, traditionally known as the local oscillator signal, is derived from the high stability 5 MHz reference oscillator in the BDE for signal mixing purposes. It is essential that this oscillator is highly frequency stable

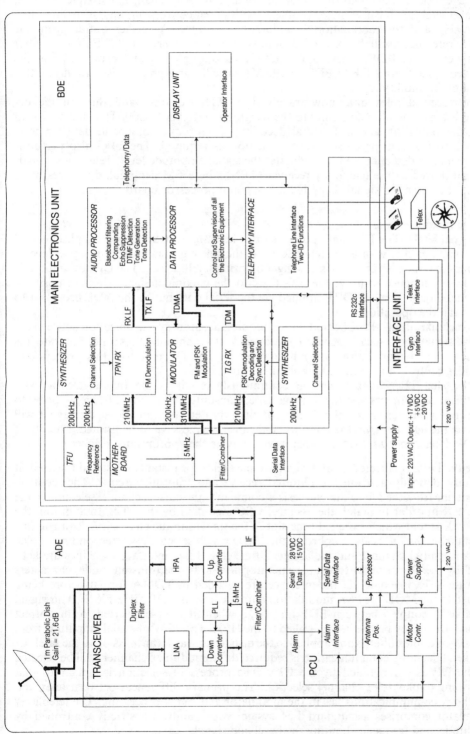

Fig. 6.16   Saturn 3S.90 SES system description. (Courtesy ABB NERA AS)

over a considerable range of operating conditions. Any change of frequency here, however slight, will cause major problems with down conversion, up-conversion, data processing and the modulation/demodulation circuitry. A further output from the 5 MHz reference oscillator is applied from the timing frequency unit (TFU) to the filter combiner in the BDE, where it is multiplexed onto the coaxial cable carrying the received signal from ADE to BDE. A 190 MHz oscillator is phase locked to this 5 MHz to maintain stability.

The received telex data now travels from ADE to BDE and through the de-multiplexer in the filter combiner to the telegraphy receiver (TLG). This unit performs PSK demodulation and differential decoding to reproduce baseband data which is applied to the data processor (DP) for action as required. The TLG also performs synchronizing detection and demodulates the signal amplitude level. Telephony signals are transferred to the telephony receiver (TPN) where FM demodulation takes place before the baseband signal is applied to the audio processor board.

Data processor actions

● **Signal level**—The DP continuously stores the assigned CSC carrier amplitude level and compares this with a reference level. If the DP registers a drop in signal level it assumes that the dish antenna is in danger of losing satellite lock. The DP now initiates a step track sequence to move the dish back to the maximum signal strength position. Command data from the DP is applied via the serial data bus to the ADE processor to command the operation.

● **Synchronization**—PSK demodulation requires accurate synchronization between the receive signal and a locally generated frequency. If synchronization is not accurate it is possible that demodulation will fail completely. The DP monitors this process and provides control to the synthesizer circuits as required.

● **Baseband TDM data**—The DP continuously monitors the Inmarsat SES IMN identification numbers and the message type, transmitted in the data frame, in order to determine what action to take. If the DP circuitry detects a broadcast message it will activate the telex unit via the interface and cause the message to be printed. Equally, if the DP detects its own IMN number it will activate the appropriate peripheral.

Once the DP has detected a valid signal (duplex telex for instance) destined for itself, it will automatically activate the required receive and transmit circuits. The 'control' output line from the DP to the synthesizer commands the correct channel selection in the TLG RX in order to match the assigned CSC identified by the NCS. In addition, the TLG RX is told which assigned time slot to search within the channel. A 'control' line output from the DP is also applied to the MOD which selects the corresponding duplex transmit paired channel and the assigned time slot. Telex interface machines are acti-vated by the DP in order to enable operator access. In the transmit mode, characters from the telex machine via the interface are encoded in TDMA format before being applied to the MOD for PSK modulation onto the nominal 310 MHz TX IF. This signal is coaxially coupled between BDE and ADE where it is up-converted to the correct channel frequency before being applied to the HPA. The 25 W of signal amplitude from the HPA is duplex filtered and applied to the dish for upward transmission to the satellite.

When telephony communication is required different duplex channels will be selected by the DP as commanded by the CES. The process of selection is similar to that describing duplex telex channel selection. The primary differences between telex and telephony communications lies in the modulator and signal processing. The telephony modulator comprises a standard FM system with bandwidth strictly controlled by amplitude, and frequency limiting the baseband signal.

Push-button used for **distress calls** only.
The **emergency indicator** goes ON when the emergency button is pressed and the main telephone is off-hook.
The **emergency indicator** also goes on if **P3** is typed on the teleprinter.

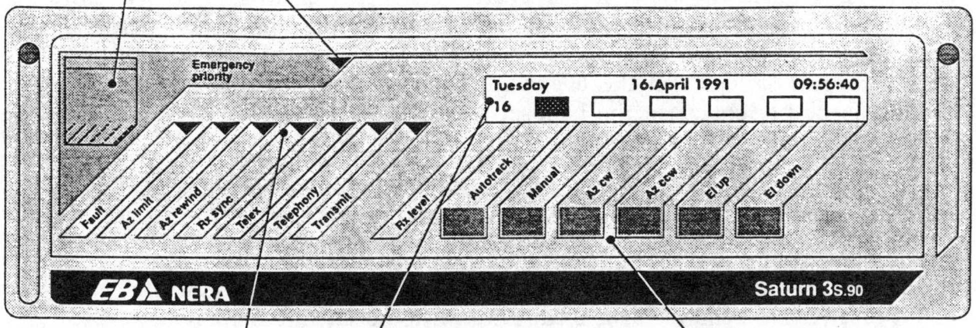

Emergency priority
push-button   indicator

Fig. 6.17 Saturn 3S.90 operator control panel

Indicators

Fault
RED flashing light indicates fault.

Az limit
RED light indicates antenna in rewind area. (See page 45.) Automatic rewind if no traffic

Az rewind
RED light indicates antenna rewinding. No traffic can be handled as long as indicator is lit.

Rx sync
GREEN light indicates Rx sync.

Telex
GREEN light indicates telex traffic.

Telephony
GREEN light indicates telephony traffic.

Transmit
GREEN light indicates that the transmitter is ON.

Display

The display provides the following indications:
First line:
- **Day, date and time** or
- **Gyro (ship's heading)**
- **Azimuth and elevation angle**

Second line:
- **Rx level**
- **Push-button indicators**

Telephone set(s)

The telephone set(s) is used for telephone calls and setting up facsimile/data communication.

Teleprinter

The teleprinter is used for

- Telex traffic
- Entering of data, such as time, gyro, AZ and EL.
- Antenna control
- Equipment tests.

Push-buttons

Autotrack
Sets the antenna to **Autotrack** mode. Time and date will be shown on the display.

Manual
Sets the antenna to **Manual** mode. Gyro, azimuth and elevation will be shown on the display.

Az cw
Turns the antenna clockwise as long as the button is pressed and the antenna is in **Manual** mode

Az ccw
Turns the antenna anti-clockwise as long as the button is pressed and the antenna is in **Manual** mode

El up
Turns the antenna upwards as long as the button is pressed and the antenna is in **Manual** mode.

El down
Turns the antenna downwards as long as the button is pressed and the antenna is in **Manual** mode.

● **Audio processor**—This performs the following baseband signal processing functions on telephony receive and transmit.

● **Signalling tone generation**—This circuit generates and detects the 2600 Hz SF tone which confirms channel selection with the CES by handshaking.

● **Baseband filtering**—This filters the audio signals from the telephone handset to the 300 to 3000 Hz band.

● **Companding**—This emphasis circuitry is not used for baseband signal processing.

When a voice channel is selected, the baseband audio signal to be transmitted is compressed prior to modulation and the received signal is expanded after demodulation. The process of compression and expansion is known as companding.

● **Echo supression**—In view of the great distances over which the RF carrier is transmitted in satellite systems it is usual for echo to be produced. In order not to aggravate the situation, echo suppression circuits are used in the interface between the voice baseband signals and the telephone handset. The circuitry suppresses echoes which may be present when the handset is connected via a two-wire or a four-wire system.

The seven light emitting diodes, on the display, shown in Figure 6.17, warn the operator of actions taking place in the unit as follows:

● a fault condition;
● the antenna has reached its AZ limit and a rewind sequence is occurring;
● the receiver has achieved synchronization with the satellite signal;
● telex or telephony traffic is in the process of being passed; and;
● the transmitter is on.

The liquid crystal display provides a very accurate day/date lock indication, the ship's heading and the AZ/EL angles of the antenna.

Six push buttons enable the antenna to be automatically or manually controlled.

In the automatic mode the antenna will search for and maintain lock on a satellite. See figure 6.13.

When manual is selected, the antenna may be moved clockwise (AZ CW) or anticlockwise (AZ CCW) in azimuth, or upwards (EL UP) or downwards (EL DOWN) in elevation, whilst the operator looks for the satellite using the tables shown in figure 6.10. The signal strength meter will indicate when a satellite has been found.

Japan Radio Company JUE-45A SES

Since commencing production in 1977, JRC has become the world's largest supplier of

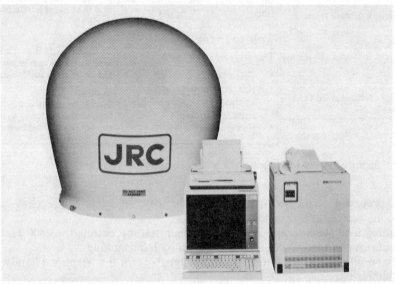

Fig. 6.18   A picture of the JRC JUE-45A Inmarsat-A SES

ship earth stations. It is highly probable therefore that readers will come into contact with a JRC JUE series SES at some point in their seagoing career.

The JUE-45A mark II has been designed to comply fully with the carriage requirements for the GMDSS. The terminal provides all the communications features required by a modern seagoing vessel. A multichannel version is available, JUE-45AM II, providing two simultaneous communication channels.

Technological details of the JUE-45A can be found in the companion volume *Satellite Communications: Principles and Applications*. A brief outline of some of the systems and operation follows.

- **Automatic antenna scanning and satellite selection**
An automatic search and recovery system enabling communications to be reliably maintained.
- **Abbreviated dialling and redialling**
A store of up to 40 abbreviated numbers and codes enables quick dialling to be made to popular shore-based subscribers.
- **A range of telephone interfaces**
These include; remote telephones, facsimile transceivers, PABX interfaces, data modems and personal computer (PC) connection.
- **Automatic reception of data transmission**
The onboard DTE can be automatically connected to a shore-based terminal unit on a voice channel without the intervention of an onboard operator.
- **Telex file transmission by automatic request**
Automatic transmission of a file to a shore subscriber can be commanded by the operator.
- **Automatic file transmission by polling command**
A shore subscriber can directly access an onboard file by use of a password.
- **Verbal instructions**
Telephones may be operated by following synthetic voice instructions (short English verbal commands) that are given from the handsets in an interactive mode.

The variety of options which may be used in the GMDSS includes the following.

- **Voice Distress Box (VDB)**
By using the VDB a voice distress call can be requested from a remote location even when the SES is being used.
- **Distress Message Unit (DMU)**
In an emergency, by pressing the two buttons on the DMU, the stored distress information (ship's name, nationality, position, nature of distress,etc) is automatically transmitted to the RCC via a telex channel.
- **Automatic ship's position reporting system**
Navigational information, including the ship's position taken from the satellite navigator, is reported to the shore office via a telex channel on receipt of a polling command. By adding the telex channel interface the SES can be pre-programmed to originate telex calls, automatically making scheduled transmission of the ship's position without the intervention of the operator.
- **EGC decoder**
By adding an EGC decoder unit the SES is able to receive SafetyNET and FleetNET messages.

The JUE-45A uses a computer terminal for centralized control of the entire SES system and to provide word-processing capabilities. A computer VDU used in this way

```
SES Status    REC. Level      EIRP        Channel Type          Warning/Error            Page

READY      RECO77    EIRP35.6   VOICE                                          SES  CONT
  -ITEM-             -STATIE    -RENEWAL-              -OPERATION GUIDANCE-
BEARING(DEG)            0                           (DISPLAY ONLY)
HEADING(DEG)          100   SET:_                    (0 TO 359: SHIP'S HEADING)
AZIMUTH(DEG)           45                            (0 TO 359)
ELEVATION(DEG)         30                            (5 TO 90)
SATELLITE             POR                            (P:POR, I::IOR, AI:AOI, A2:AO2)
TDM FREQ NO            0                              (0 OR I)
PRIORITY           ROUTINE                           (R:ROUTINE, S:SAFETY, U:URGENT)
NETWORK                I                             (I TO 7)
OCEAN AREA            I5                             (0 TO 7, I0 TO I7)
SHORE ID             02                              (I TO 7, I0 TO I7)
TELEPHONE             I                              (I TO 3)
EXT.  BUZ        I:   ON                             (ON:ON, OFF:OFF)
                 2:  OFF                             (ON:ON, OFF:OFF)
VDU              MASTER                              (M:MASTER, S:SLAVE)
DATE           I2/0I/I990                            (MONTH/DAY/YEAR)
TIME             I3:30                               (HOUR:MINUTE)

ROUTINE      AREAI5   CESO2 POR TELI   MASTER-VDU                         I2/0I  I3:30
 Priority    Ocean Area  CES ID/SAT  TEL   VDU           Information           UT/Timer
```

Fig. 6.19   The JUE-45A Inmarsat-A SES VDU control display. (Courtesy JRC)

```
SES Status    REC. Level      EIRP        Channel Type          Warning/Error            Page

READY      RECO85    EIRP36.0   TELEX                                          FLIST PI
....+....I....+....2....+....3....+........+........5....+.........
NO         FILE NAME     VOLUME           NO         FILE NAME     VOLUME
 0         RX/TX TEXT      I234            I6         QBF MSG          86
 I         TEXT  0I        2058            I7   R     JRC             I88
 2         TEXT  02       I029
 3         TELEXO8/I0      876
 4   R     TELEXO8/II      5I7
 5   R     TELEXO8/I2      628
 6         EXP   0I        226
 7         EXP   02        3II
 8         EXP   03        404
 9         EXP   04        236
I0         EXP   05       I25
II   R     83.03/03TX      7I2
I2         83.04/I2RX      583
I3         83.05/05TX      328
I4         83.06/20RX      4I2
I5   R     83.07/07TX      635
FILE NO =

RESIDUE= 6524CHAR                        MASTER-VDU                         I576SEC
 Residual Capacity                           VDU           Information           UT/Timer
```

Fig. 6.20   The JUE-45A SES FILELIST display. (Courtesy JRC/Raytheon)

is a very flexible method of displaying a multitude of data. As an example see figures 6.19 and 6.20.

The two figures shown are of the VDU information display in the SES Control mode and the File LIST mode.

The SES Control mode essentially displays system status and would be the normal display when the word processor is not being used. Main system parameters are displayed on the top and bottom lines of the screen.

SES STATUS INDICATION		
Indication	SES State	Remarks
READY	Ready State	Ready for communication
QUEUE	Queue State	Received Queue Message from CES
BUSY	Busy State	Received Congestion Message from CES
ERROR	Error State	Received Request Not Acceptable Message from CES
BDU? (Inversion)	BDU Error State	Error in Data communication with BDU
No indication	Not Ready State	Not ready for communication

Fig. 6.21   SES STATUS INDICATION. (Courtesy JRC/Raytheon)

• **SES STATUS.** This part of the screen shows one of six indications as listed in the table. An operator can readily see the state of the SES equipment.

• **REC LEVEL.** Indicates the receiving signal level of the assigned CSC frequency transmitted from the NCS. The indication range value is from 000 to 127 with a nominal value of approximately 80. This enables an experienced operator to judge whether the communications link is satisfactory or whether a new satellite should be accessed.

• **EIRP.** Shows the EIRP of the transmitted carrier from the SES. EIRP indication range value is from 33 dBW to 40 dBW in 0.2 dB increments. The EIRP indications are shown in figure 6.22.

EIRP INDICATION	
Indication	EIRP Condition
EIRP 35.6	EIRP × 35.6 dBW
EIRP–LT33	25 to 32.9 dBW
EIRP–GT40	More than 40.1 dBW
EIRP 0	0 to 24.9 dBW
EIRP 0	Power to PA OFF
SPACE	Carrier OFF

Fig. 6.22   EIRP INDICATION. (Courtesy JRC/Raytheon)

CHANNEL TYPE INDICATION	
Indication	Channel Type
TELEX	Duplex telegraphy.
VOICE	Duplex telephony with compandors.
D/FAX	Duplex telephony without compandors.
TELEX (RX)	Shore-to-ship one-way telegraphy.
VOICE (RX)	Shore-to-ship one-way telephony with compandors.
D/FAX (RX)	Shore-to-ship one-way telephony without compandors.
TELEX (RX): VOICE	Shore-to-ship one-way telegraphy and duplex telephony with compandors.
TELEX (RX): D/FAX	Shore-to-ship one-way telegraphy and duplex telephony without compandors.
HSD	Ship-to-shore High Speed Data (Option)

Fig. 6.23   CHANNEL TYPE INDICATION. (Courtesy JRC/Raytheon)

WARNING INDICATIONS		
Indication	Warning/Error State	Remarks
TELEX MEM.OVF	File O overflow	
EDT MEM.OVF	Edit file overflow	Buzzer sounds for 150 ms
TLX, EDIT M.OVF	File O and edit file overflow	Ditto
FILE NO.OVF	File number overflow	Buzzer sounds for 100 ms
SAME NAME	Same file name	Ditto
FILE NO.?	Incorrect input of file number	Ditto
FILE NAME?	Incorrect input of file name	Ditto

Fig. 6.24   WARNING INDICATION. (Courtesy JRC/Raytheon)

PAGE INDICATION	
Indication	Name of Page
ON LINE	ON LINE page (file 0)
SES CONT	SES CONT page
FLIST P1~P5	FILE LIST page
EDIT	EDIT page (File 1 – 31)
LOCAL	LOCAL page
IDLE	IDLE page or LOCAL TCI page

Fig. 6.25 PAGE INDICATION. (Courtesy JRC/Raytheon)

PRIORITY INDICATION		
Indication	Communication Type	Remarks
ROUTINE	Normal Comm.	
SAFETY	Safety Comm.	Inversion display
URGENCY	Urgent Comm.	Inversion display
DISTRESS	Distress Comm.	Inversion display & Buzzer sounds

Fig. 6.26 PRIORITY INDICATION. (Courtesy JRC/Raytheon)

- **CHANNEL TYPE.** This area indicates the type of communication channel being used.
- **WARNING/ERROR.** Provides system error or warning information. During normal operation this area remains blank.
- **PAGE.** Indicates which page is currently being displayed on the VDU screen.
- **FILE NAME.** This area indicates the displayed file name of the edit file. (Used in the EDIT mode only.)
- **PRIORITY.** Shows which communication priority has been selected. Warnings are provided for the operator on all selected modes of communication other than Routine.
- **OCEAN AREA.** Indicates the selected ocean area.
- **CESID/SAT.** Shows the selected CES ID (IMN) and the satellite ocean region.

- **TEL.** Indicates which of the on-board telephones is being used for communication.
- **VDU.** Indicates whether the master VDU or the slave VDU is being used.
- **INFORMATION.** Shows one of ten system messages as shown in figure 6.27. When two situations occur at the same time the highest priority information is displayed.

INFORMATION FROM MAIN UNIT		
Priority	Indication	Information
1	TX ALARM	Abnormal transmission of the signal burst has occurred.
2	FTU ALARM	Frequency generator of the FTU is defective.
3	SOS TELEX	Distress duplex telex communication in progress.
4	SOS VOICE	Distress duplex telephone communication in progress.
5	A/D CONV	A/D converter of ACU and MAIN is defective.
6	PA/SERVO SW	PA/SERVO switch in the ADE is switched off.
7	AUTO REWIND	Antenna is rewinding automatically.
8	ZONE	Antenna bearing is in the area of the automatic rewind ($+/- 250°$ through $+/- 270°$).
9	WAIT	Under antenna control.
10	SES CONT	SES control parameters (heading, azimuth, satellite and shore ID) are erased.

Fig. 6.27 INFORMATION FROM MAIN UNIT. (Courtesy JRC/Raytheon)

- **UT/TIMER.** This two-function display normally indicates Universal Time (UT) switching to an accurate clock timer when assignment from the CES is received. UT is shown in standard notation; Month/Day Hour/Minute, whereas the timer runs from zero to a maximum of 9999 seconds.
- **RESIDUAL CAPACITY.** Indicates the remaining capacity of the edit file in the range zero to 16384 characters. This figure assists the operator during the preparation and storing of messages.
- **FILE NAME.** Shows the titles of stored files.

- **VOLUME.** An indication of the size of a file quoted in characters.

Inmarsat-A transportable equipment

A number of companies manufacture Inmarsat-A transportable equipment which, in effect, is a take-anywhere 'satellite phone'. The major news gathering services of the world are the premier users of this system. Instantaneous on-scene reporting of major news items directly into the news wire services has now become commonplace using transportable equipment.

Weighing between 20 and 50 kg, the equipment is self-contained in one or two suitcases, which contain a foldaway parabolic antenna, the communications electronics and the telephone or computer terminal. Such equipment must conform to Inmarsat-A specifications and is able to provide all Inmarsat-A services except priority 3 distress alerting.

The two main features which make this type of equipment different from an Inmarsat-A SES installation are the power supply and the antenna arrangement.

Power supply

When used without the aid of a generator the transportable unit will operate from 24 V lead acid batteries. Power consumption on receive is approximately 75 W whereas on transmit it doubles to approximately 150 W, leading to a current drain of approximately 6.25 A. However, it should be remembered that the transmitter is not on the air continuously so the continuous current drain will not be that high. Also, a number of power saving systems are used such as automatic power-down.

Antenna

As with all Inmarsat-A land earth stations (LESs) the equipment is dependent upon antenna gain for both transmit and receive, and consequently uses a parabolic dish antenna. The antenna, between 0.8 and 1 m in diameter, is sectionalized or folded to fit in the carrying case. It must therefore be assembled and manually aligned with the satellite. This is done using charts similar to those produced by ABB NERA AS Communications (figure 6.10) or by following the simple routine produced by Inmarsat.

Mobile antenna alignment

The AZ/EL value to be used when aligning the antenna may be derived as follows.

- Determine the LES position relative to the satellite.
- Find the difference in longitude between LES and satellite (e.g. if the satellite longitude is 15.5° west (AORE) and the LES longitude is 15.5° east the difference is 31°.)
- Round off the longitude to the nearest 5°.
- Round off the latitude of the LES to the nearest 5°.
- From the LES position, calculate in which quadrant the LES lies relative to the orbital location of the satellite.
- Read-off the AZ and EL figures from the relevant quadrant table. There are four published tables for when the LES is located NW, SW, NE or SE of the satellite. The NW table is shown in figure 6.28.

LATITUDE (DEGREES NORTH)	DIFFERENCE IN DEGREES LONGITUDE BETWEEN SHIP AND SATELLITE																
	0	5	10	15	20	25	30	35	40	45	50	55	60	65	70	75	80
0	180	090	090	090	090	090	090	090	090	090	090	090	090	090	090	090	090
	90	84	78	72	66	61	55	49	44	38	33	28	22	17	12	07	02
5	180	135	116	108	103	101	099	097	096	095	094	093	093	092	092	091	091
	85	82	77	71	66	60	55	49	43	38	32	27	22	17	11	06	01
10	180	153	135	123	116	110	107	104	102	100	098	097	096	095	094	093	092
	78	77	73	69	64	59	53	48	43	37	32	27	21	16	11	06	01
15	180	161	146	134	125	119	114	110	107	105	102	100	098	097	095	094	093
	72	71	69	65	61	56	51	46	41	36	31	26	21	16	11	06	01
20	180	166	153	142	133	126	121	116	112	109	106	103	101	099	097	095	093
	66	66	64	61	57	53	49	44	39	34	30	25	20	15	10	05	01
25	180	168	157	148	139	132	126	121	117	113	110	106	104	101	099	096	094
	61	60	59	56	53	50	46	41	37	33	28	23	19	14	09	05	00
30	180	170	161	152	144	137	131	126	121	117	113	109	106	103	100	098	095
	55	54	53	51	49	46	42	38	34	30	26	22	17	13	09	04	00
35	180	171	163	155	148	141	135	129	124	120	116	112	108	105	102	099	-
	49	49	46	46	44	41	38	35	31	28	24	20	16	12	08	04	-
40	180	172	165	157	150	144	138	133	127	123	118	114	110	107	103	100	-
	44	43	43	41	38	37	34	31	28	25	21	18	14	10	07	03	-
45	180	173	166	159	153	147	140	135	130	125	121	116	112	108	104	101	-
	38	38	37	36	34	33	30	28	25	22	19	16	12	09	05	02	-
50	180	173	167	161	155	149	143	138	132	127	123	118	114	110	106	102	-
	33	32	32	31	30	28	26	24	21	19	16	13	10	07	04	01	-
55	180	174	168	162	156	150	145	139	134	129	125	120	115	111	107	102	-
	27	27	27	26	25	23	22	20	18	16	13	11	08	05	03	00	-
60	180	174	168	163	157	152	146	141	136	131	126	121	117	112	107	-	-
	22	22	21	21	20	19	17	16	14	12	10	08	06	04	01	-	-
65	180	174	169	164	158	153	148	142	137	132	127	122	118	113	-	-	-
	17	17	16	16	15	14	13	12	10	09	07	05	04	02	-	-	-
70	180	175	169	164	159	154	148	143	138	133	128	123	118	114	-	-	-
	11	11	11	11	10	09	09	07	07	05	04	03	01	00	-	-	-
75	180	175	169	164	159	154	149	144	139	134	129	124	-	-	-	-	-
	06	06	06	06	05	05	04	04	03	02	01	00	-	-	-	-	-
80	180	175	170	165	160	155	150	-	-	-	-	-	-	-	-	-	-
	01	01	01	01	01	00	00	-	-	-	-	-	-	-	-	-	-

For any calculation the table is arranged with the Azimuth reading above the Elevation reading e.g.   AZM
ELN

Fig. 6.28   AZ/EL reading for a vessel situated North and West of the satellite. (Courtesy Inmarsat)

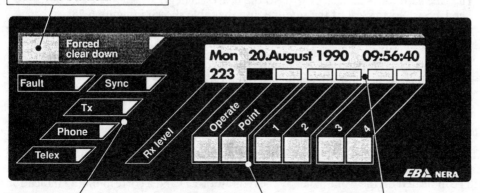

Forced clear down
Used if normal release for telephone (on-hook) or telex (off-line) does not clear the station.

Indicators

Fault
RED flashing light indicates fault.

Sync
GREEN light indicates Rx synch.

Telex
GREEN light indicates telex traffic.

Phone
GREEN light indicates telephony traffic.

Tx
GREEN light indicates that the transmitter is ON.

Pushbuttons

Operate
Sets the CompacT to operating mode. Date and time is shown on the display.

Point
The antenna can be repositioned. Adjust to obtain maximum Rx level on the display.

1, 2, 3, 4
For testing only.

Display

The following may be displayed:
First line:
- **Day, date and time**
- **Rx level indicator**
- **Status/alarms indications**

Second line:
- **Rx level (3 digits)**
- **Pushbutton indicators**

Telephone set(s)

The telephone set(s) is used for telephone calls and facsimile/data communication.

Teleprinter

The teleprinter is used for

- Telex traffic
- Equipment tests

Fig. 6.29 Saturn compact-T operator control panel. (Courtesy EB NERA)

Example

LES Latitude	23 deg.45 min. North (25 deg. North)
Longitude	139 deg.11 min. East (139 deg. East)
Satellite longitude	180 deg. East (POR)
Difference in longitude	180 deg. – 139 deg. = 41 deg. (40 deg. to nearest 5 deg.)

The LES latitude is 25 deg. North (to the nearest 5 deg.)

The LES is NorthWest of the satellite, use the table shown in figure 6.28.

At latitude 25 deg. North and longitude 40 deg. West the AZ angle = 117 deg. and the EL angle = 37 deg.

If the transportable equipment is mounted in a vehicle it is possible to use dynamically driven antennae which automatically track the satellite once aligned.

Several companies offer transportable Inmarsat-A terminals. See Appendix 3.

The Saturn Inmarsat-A Compact-T

After the antenna is assembled and the power connected, the unit runs through a diagnostic check as part of the 'system startup' procedure. Any error which is diagnosed will be indicated on the front panel by the red fault LED and a corresponding indication on the display. When the system is successfully initialized, it enters the operate mode, shown in figure 6.29.

The antenna is now aligned by pressing the 'point' command and using AZ/EL figures derived from the charts shown in figure 6.10. The signal strength is indicated on the display.

The unit may now be used for telephone/telex traffic using the standard Inmarsat-A calling sequences.

7
The Inmarsat-B system

7.1 Introduction

The Inmarsat-B system has been designed to become the ultimate successor to the ageing Inmarsat-A system for the provision of mainstream professional mobile communication services into the next century.

Inmarsat-A system technology is now ageing. The system has been expanded to the effective limit of its analogue technology. The number of Inmarsat-A users has grown dramatically as have their needs. Inmarsat-A is supporting far more traffic than was intended in the original design and consequently a change to new digital technology, as utilized in the Inmarsat-B system, is necessary. Continuing system development is driven by consumer demand, e.g. better system access, better quality and more sophisticated services amongst other things. These factors in no way detract from the capabilities of the Inmarsat-A system which has been the workhorse of mobile satellite communications for nearly two decades. Indeed, Inmarsat-A will continue to provide communications services, in parallel with Inmarsat-B, until at least the year 2005 when it will be totally replaced by the new system.

The new, fully digital Inmarsat-B system became operational in late 1993 to provide global mainstream satellite communications between fixed stations (LES) and mobile terminals (MES).

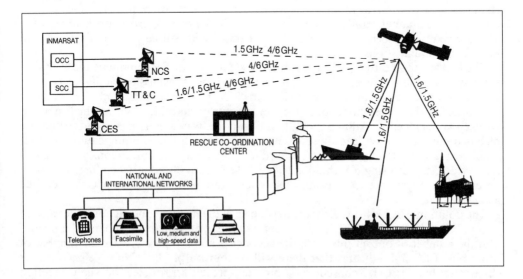

Fig. 7.1   The Inmarsat-B network configuration

Inmarsat-B coverage, performance, availability and SES environmental conditions are identical to Inmarsat-A and are fully compatible with IMO requirements for distress operation within the GMDSS radionet. It should be noted however, that the technology of Inmarsat-B is far superior to that used in the Inmarsat-A system and, consequently, the two systems are not technologically compatible

Outline system description

Inmarsat-B system design is based on the very latest digital systems technology. This technology has been used extensively in order to reduce satellite channel requirements, when compared with the Inmarsat-A system, by approximately 50%. The use of digital coding and modulation techniques, voice-activated carrier switching and forward power control on the SCPC forward communications carriers has led to both bandwidth and power reductions of this order. The extensive use of phase shift keying, offset-quadrature PSK modulation on the SCPC channels, bi-phase BPSK modulation on the TDM channels and FEC, enables these savings to be made. As an example the voice channel bandwidth has been reduced from 50 kHz to 20 kHz and the typical transmit power output from a MES has been reduced from 40 W to 25 W when compared with the Inmarsat-A system. See also Chapter 4.

Inmarsat-A voice baseband processing using FM has been replaced with digital voice channel coding. The use of digital speech processing enables large bandwidth and power savings to be made whilst maintaining the excellent quality produced by Inmarsat-A MESs. A combination of all these factors plus the use of unpaired transmit and receive channel frequencies leads to a considerable improvement in the use of the frequency spectrum available.

Inmarsat-B has available for use a much greater frequency spectrum band of 20 MHz which is almost three times the space available to Inmarsat-A operations at 7.5 MHz.

Spectrum efficiency will be further improved with the use of the spot beam communication facilities to be used on the new Inmarsat INM3 satellites. Instead of using the one-third earth coverage transponder, two ships in the same ocean region may be able to use spot beam transponders with a much narrower beamwidth enabling simultaneous channel re-use to be made in the same ocean region. The return request channel and the telex message channel detailed above refer to 1/2 FEC. Forward error correction (FEC), as described in the terrestrial telex section of this book, is used in a modified form in satellite communications.

The FEC method differs from the method used for terrestrial communications in that the system is able to correct any errors detected in the transmission path. In common with all error detection and correction systems FEC requires that a number of data bits are added to the information bits to form an appropriate bit pattern which may be checked at the receiver. Adding redundant bits in this way greatly increases the number of bits to be transmitted leading to greater inefficiency in the communications system. FEC used in the satellite link uses a procedure which operates continuously on the bits in the data stream using a technique called Viterbi convolutional encoding. It is only necessary to appreciate that, using this system, different levels of correction can be achieved.

The notation FEC 3/4, 1/2 etc indicates the level of encoding and signifies the ratio of output bits to input bits. FEC 1/2 encoding, as used in an Inmarsat-B channel, indicates that the number of output bits from the encoder is approximately twice the number of input bits. FEC 3/4 indicates that there will be four output bits from the encoder for three input bits. FEC 1/2 consequently provides greater error detection and correction than FEC 3/4.

The space segment

Inmarsat-B uses the same space segment as all the other Inmarsat systems. The new INM2 satellites carry a number of transponders, one of which is dedicated to Inmarsat-B operation. The next generation of satellites, INM3, currently under development, will also carry spot beam facilities for use with the Inmarsat-B system.

The use of spot beams is another significant difference between the Inmarsat A and B systems. Their use will lead to greater economy of operation in regions of high traffic density where channel re-use in an ocean region is possible

Inmarsat-B is able to support the new Inmarsat high-speed data services and, consequently, is fully compatible with the CCITT integrated services digital network (ISDN).

The ground segment

Inmarsat-B utilizes the same ground control, access and communications station network as described earlier. Most of the existing NCSs and LESs currently providing Inmarsat-A services have refitted signalling, access and control equipment in order to operate within the new technology used in Inmarsat-B. Enhanced user services are offered at the discretion of the telecommunications company owning the LES and thus will differ. A number of Inmarsat-B LESs exist in each ocean region to provide the fixed end of the communications link. It is the responsibility of these LESs to assign the TDM/TDMA channel (data communications).

As with other Inmarsat systems, the NCS in each ocean region plays a vital role in the operation of the system by co-ordinating access to the communications channels. With reference to figure 7.2 it can be seen that the NCS is provided with signalling links with both the SES and the CES.

The NCS is responsible for the assignment of communication channels for SCPC (voice) communications. In addition, the NCS monitors all access, control and signalling channels and maintains a database on MES status etc.

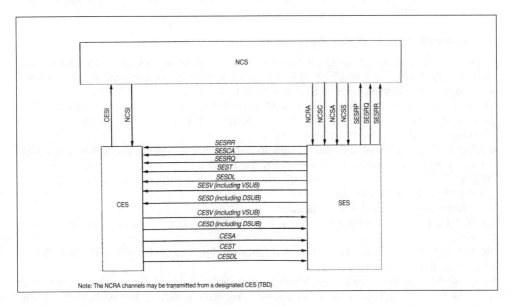

Fig. 7.2 Inmarsat-B system functional channel configuration

Distress requests and replies are monitored by the NCS which is able to take executive action should no reply to a distress call from an SES be forthcoming.

Ship earth stations

Inmarsat-B SESs are similar in design to those currently in use in the Inmarsat-A system. The equipment operates in the transmit band 1626.500 to 1646.500 MHz and in the receive band 1525.000 to 1545.000 MHz. This wider band of frequencies coupled with reduced bandwidth requirements and the introduction of new technology essentially leads to more channels being available. Also, channels are unpaired and are assigned on a demand basis from a pool of frequencies, leading to very efficient management of the limited spectrum available.

Typical ADE equipment is very similar in appearance and size to that used in the Inmarsat-A system. This includes a parabolic antenna of approximately 0.9 m in diameter, inside a protective radome, with full stabilization and satellite tracking assemblies.

As with the Inmarsat-A system, the antenna assembly is fully stabilized in order to satisfy the IMO regulations for compliance with the SOLAS carriage requirements for vessels operating within the GMDSS radionet.

Vessel motion specifications include, $\pm$ 30° roll, $\pm$ 10° pitch, 6° turning rate and 1 deg s^{-2} angular acceleration rate.

The use of the same antenna and stabilization specifications of the Inmarsat-A system enable Inmarsat-B equipment manufacturers to reduce development costs significantly by the re-use of major portions of existing designs.

With the exception of voice communication, the Inmarsat-B BDE equipment closely resembles the design of an Inmarsat-C terminal based on a personal computer.

There are two classes of Inmarsat-B SES as follows:

- Class 1 SES telephone and telex services,
- Class 2 SES, telephony only (plus the capability to receive Inmarsat service announcements).

System access control and signalling

To achieve and maintain good control of the system access and communications channels the NCS, CES and SES utilize various signalling channels as detailed below.
- **Assignment channels CES to SES and SES to CES**
SESRQ (SES request channel). Carries SES signalling information to the selected CES. Specifically used for the access main request and acknowledgement messages. Monitored by the NCS for distress purposes.
CESA (CES assignment channel). Used to carry CES signalling messages to an SES. Specially used for channel assignments for telex calls.
NCSA (NCS assignment channel). Carries channel assignment messages to SESs for voice communication channels.
- **Information and monitoring channels**
CESI (CES inter station channel). Used by each CES to carry signalling information to the NCS.
NCSI (NCS inter station channel). Used from the NCS to each CES in the network to carry signalling information.
NCSC (NCS common signalling channel). Carries NCS signalling messages from shore to SESs for call announcement broadcasts, selective clearing, network status information and bulletin board.

SESRP (SES response channel). Used to carry SES signalling information to the NCS. Specifically, the response information required for shore originated calls, and for the acknowledgement of SES group ID down-loading messages.

NCSS (NCS spot beam channel). To be used in the future, in the C-L direction (one frequency per spot beam) to enable SESs to identify the satellite spot beam in which they are located. INM3 satellites will carry spot beam facilities.

● **SCPC communication channels. (Access controlled by the NCS)**

SESV (SES voice) and **CESV** (CES voice). Digital voice channels. Used in both the forward (to ship) and reverse (from ship) directions.

The channels contain sub-band signalling information **(VSUB)** for channel control during the communications link.

SESD (SES data) and **CESD** (CES data). Data channels with similar usage and sub-band signalling to voice channels. The forward data channel is labelled CESD and the return channel SESD. Sub-band signalling is labelled **DSUB**.

● **TDMA/TDM communications channels. (Access controlled by the CES)**

SEST (SES telex channel). Carries telex to a CES.

CEST (CES telex channel). Carries telex to the SES.

SESDL (SES low-speed data channel). Used to carry low-speed data to a CES.

CESDL (CES low-speed data channel). Carries low-speed data to a SES. Same rate as for SESDL.

Figure 7.2 illustrates the Inmarsat-B system functional channel configuration.

Call set-up procedures

Duplex radio-telephony from the mobile

The set-up sequence is initiated by a call, which includes the following information, from the SES to the selected CES on the SESRQ channel. The *access-request message* includes the following data:

CES identification.

SES identification.

Request service type. (Telephony, telex or data.)

Priority. (Normal or priority 3, maritime SES only.)

SES antenna azimuth zone (for frequency assignment purposes).

SES antenna azimuth angle (distress priority only).

SES antenna elevation angle (distress priority only).

Satellite spot beam identification (for future use).

See figure 7.3 which illustrates the action sequence between the signalling units (SU).

Assuming the *access-request message* carries normal priority, the CES returns a *request-for-channel-assignment message* using the CESI channel to the NCS. The NCS will respond to the SES over the NCSA channel, by sending a *SCPC-channel-assignment message* and, using the NCSI, informs the CES of the channel assignment. The SES identity is then included in the network SES busy list at the NCS and the local SES busy list at the CES.

Note: when the new generation of INM3 satellites is in use, the NCS will attempt to assign a spot-beam channel frequency. If one is not available, or until the time when INM3 satellites are in use, a global-beam channel is assigned. This sequence establishes the communication channel link.

The SES now transmits the called-party identification to the CES via the forward sub-band channel (VSUB) using the *service-address message* along with a *scrambling-vector message* to ensure call security is maintained. The messages are sent continuously

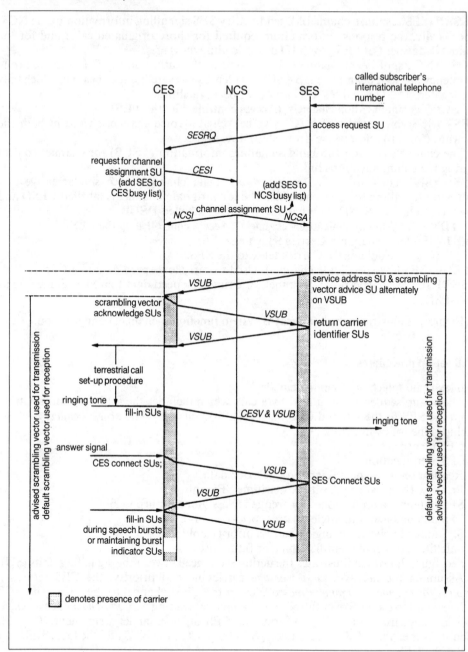

Fig. 7.3   Ship-originated duplex telephone call set-up sequence

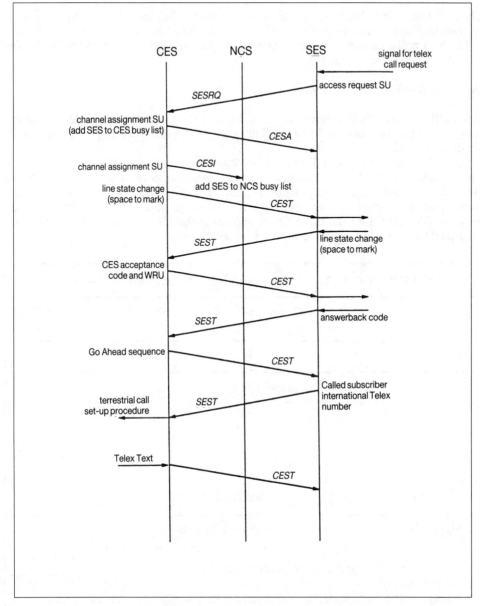

Fig. 7.4 Ship-originated duplex telex call set-up sequence

until acknowledged by the CES using the *scrambling-vector-acknowledgement message* on the forward sub-band channel VSUB. The duplex voice channel now starts to carry traffic.

Duplex radio-telephony from shore
A CES receives, via landline from the Maritime Satellite Switching Centre, the number of the SES to be called. A 'T' digit is used to indicate that the Inmarsat-B system is to be used for the call. The CES now sends a *request-for-call-announcement message* via the

CESI channel to the NCS. If the SES identification is not included on the busy list a *call-announcement message* is sent on the NCSC to the SES.

The SES replies using a *response message* on the SESRP channel to the NCS. The procedure then continues as before.

Duplex radio-telegraphy from the mobile (telex)

Telex call set-up initially follows the same lines as those outlined for voice communications with the transmission by the SES of an *access-request message* to the CES on the SESRQ channel.

Telex communication uses TDMA/TDM channels which are assigned by the CES. Consequently, the CES now transmits a *channel-assignment message* to the SES using the CESA channel. This message includes the communications channel to be used plus the time-slot assignment information. The SES identity is now added to the CES local busy list. The CES informs the NCS by means of the CESI channel of the channel-assignment and the SES is added to the network busy list.

INMARSAT-B TDM/TDMA CHANNEL ASSIGNMENT Frame S7								
BIT number								OCTET No
8	7	6	5	4	3	2	1	
Message type								1
SES ID								2
								3
								4
Spare			SES transmit time slot					5
SES EIRP		SES receive time slot						6
SES receive channel number								7
								8
SES transmit channel number								9
								10
CCITT CRC								11
								12

(Courtesy Inmarsat)

Fig. 7.5 Request and assignment data frames

INMARSAT-B ACCESS REQUEST (NON-DISTRESS) Frame S3								OCTET No
BIT number								
8	7	6	5	4	3	2	1	
Message type								1
SES ID								2
								3
								4
CES ID								5
Azimuth angle zone			Priority		Elevation angle zone			6
Service nature		Service type						7
Channel parameters								8
Terrestrial network ID								9
init/ repeat	spare	Spot beam ID						10
CCITT CRC								11
								12

(Courtesy Inmarsat)

Fig. 7.5   (cont.)

8
The Inmarsat-C system

8.1 Introduction

Inmarsat-C is an all-digital text/data system which operates between the mobile earth station (MES) and a 'gateway' or land earth station (LES). The principal advantages of the system when compared with Inmarsat-A, are that it is low cost, small and uses a very small omni-directional antenna. Its main disadvantage is, of course, that voice communication is not possible. It should be remembered however, that information transfer by voice communication is very slow and generally requires more bandwidth per carrier than data communications.

This go-anywhere system became fully operational, after a long period of pre-operational trials, in January 1991 and has been further developed since that date.

Inmarsat-C mobile terminals generally use a laptop computer as the operator interface,

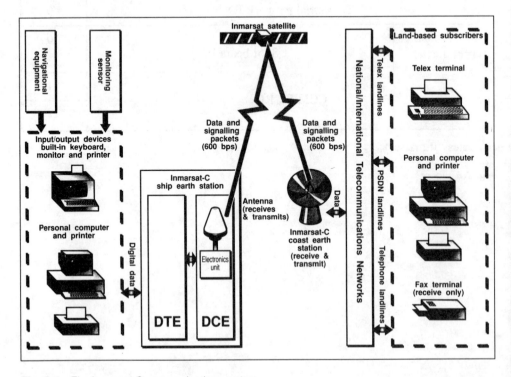

Fig. 8.1 The Inmarsat-C communications system

which becomes the data terminal equipment (DTE), whereas the interface between this unit and the satellite is known as the data circuit terminating equipment (DCE).

Outline system description

Inmarsat-C terminal's store-and-forward feature enables the system to interface with any terrestrial message or data networks including telex, X.25, X.400 and voice band data networks via the public service telephone network (PSTN) or the integrated services digital network (ISDN).

Data is transferred between MES and LES at a rate of 600 bits s^{-1} for second generation satellites, and at 300 bits s^{-1} for first generation satellites, in a 'packetized' stream which is formed of interleaved 8.64 second frames providing 10,000 frames per day. The relatively slow rate of data communications coupled with an efficient error detection and correction system ensures that even when signal fading exceeds one second—an exceedingly long duration in a data communications system—there will be no data packet loss detected by the receiving operator.

The space segment

The Inmarsat-C service was developed to use a mixture of satellites, each with varying characteristics. First generation satellites, MARECS and INTELSAT-V MCS, provide lower transponder gains than the second generation INM2 satellites. This, in turn, leads to the need for a reduction in data transmission speeds, from MES to LES, from 600 bits s^{-1} to 300 bits s^{-1} when using first generation satellites. The change is fully automatic and is based on information, received on the NCS common channel, via the NCS bulletin board.

Each of the four satellite ocean regions is supported in the same way as those described for Inmarsat-A operation, which of course uses the same satellites.

The ground segment

The network control centre (NCC) at the Inmarsat headquarters in London, supervises the four interactive NCSs using interstation signalling links, ISL. The four NCSs are as follows; the AORE and AORW NCS is situated at Goonhilly Down in the UK, the IOR NCS is at Thermopylae in Greece and the POR NCS is at Singapore. Unlike Inmarsat-A NCSs the stations communicate using ISLs and, consequently, construct an updated scenario of the system and its active status. NCSs communicate with both MESs and each LES operating in an ocean region.

LESs provide the gateway through which traffic flows between the satellite and land-based communications networks.

Currently, there are 12 LESs operating in the system with a further ten planned for the near future

Access control and signalling

A number of different types of channel are used in the Inmarsat-C system. Using the channels, information is transferred in data packets—each of which contains a check-sum permitting ARQ error detection and correction.

Channels are allocated in 5 kHz steps within the bands TX—1626.500 to 1646.500 MHz and RX—1530.000 to 1545.000 MHz. Modulation is BPSK at 1200 symbols s^{-1} (second generation satellites). FEC 1/2 rate coding is used and, con-

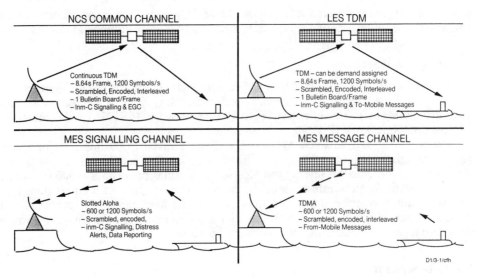

Fig. 8.2   Inmarsat-C channels

sequently, the information rate is 600 bits s^{-1} (300 bits s^{-1} for the return channel on first generation satellites). See also Chapter 4.

Signalling channels

MES signalling channels. Used during the call set-up phase. These channels are used by the MES to transmit signalling packets of information to both the NCS and the LES. Each LES has one or more signalling channels assigned to it and, in addition, there is a signalling channel associated with each NCS common channel. The channels are used, amongst other things, for requesting a message channel assignment from the LES and to the NCS for log-in purposes

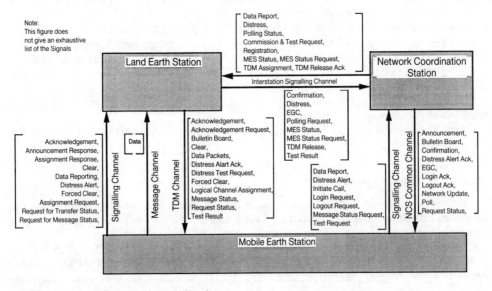

Fig. 8.3   Inmarsat-C channels and signals

Fixed station signalling channels
NCS – NCS signalling channel. Interstation signalling channels which carry system information between ocean regions.
NCS – LES signalling channel. Another channel carrying system information between the NCS in an ocean region and all its attendant LESs.

TDM channels
NCS common channel. A continuously transmitted TDM channel to which all MESs, in a particular ocean region, are tuned when in the idle state. The channel carries the important bulletin board packet which contains the information required by an MES to gain access to the system. It contains information on the static operational parameters of both the NCS and LES which are transmitting the TDM channel. Access to the channel is on a first-come first-served basis for packets of the same priority.
   There are three levels of priority:
- Inmarsat-C call announcement, polling, EGC distress priority messages, distress alert acknowledgement;
- Inmarsat-C signalling; and
- other EGC messages.
LES TDM channel. This channel has the same structure as the NCS common channel and contains, amongst other information, call set-up signalling. Access to the channel is on a first-come first-served priority basis. There are four levels of priority:
- distress priority packets;
- logical channel assignments;
- other protocol packets; and
- messages.

   TDM channel information can be displayed on the VDU of an Inmarsat-C SES. This information may be as follows.
- Frame number—The number of the current frame in use. Each 24 hour period is divided into 10,000 frames (8.64 seconds per frame) which are reset to 0000 each midnight UTC and advance in number throughout the day.
- TDM channel number—The channel number represents the frequency of the TDM channel.
- Bulletin board—This is a unique packet of information which indicates the operational status of the NCS and LES being used. The slot also contains the current frame number. Reception of consecutive bulletin board packets is used by the receiver to measure the quality of the reception. This is displayed as the 'bulletin board error rate' which, surprisingly, should be a low value. Zero is the ideal state.

Communication channels
Message channel. Used for the transfer of traffic. LES message channels are used by the MES to send store-and-forward messages to an LES. Because more than one MES may be sharing the same message channel, the destination LES allocates a transmission time to each MES. Once assigned a start time, the MES transmits all of its message without interruption. Message channels are allocated by the NCS to an LES depending upon the amount of traffic to be handled. Consequently, each LES may have more than one message channel to use

Logging-in to a NCS

Before the system can be used it must be synchronized to the NCS common signalling channel for the ocean region where the MES is situated. Synchronization is achieved by

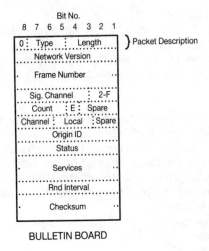

BULLETIN BOARD

SIGNALLING CHANNEL
DESCRIPTOR PACKET
- 1st GENERATION

SIGNALLING CHANNEL
DESCRIPTOR PACKET
- 2nd GENERATION

Fig. 8.4 Bulletin board signal channel descriptor frames. (Courtesy Inmarsat)

logging-in, where the MES sends a specific signal to the NCS informing it that it is now operational. If this is the first log-in for the MES the terminal automatically performs a 'Performance Verification Test (PVT)' to confirm that the unit is operational. Alternatively, a PVT may be commanded manually by the operator to verify system performance. Figure 8.5 illustrates system action during a PVT.

PVT test results are stored for later retrieval to enable the operator to establish the quality of the link.

In some cases, logging-in is automatic when the equipment is powered up and occurs at least once in every 24 hours. If the mobile is travelling within an overlap zone of two or more ocean regions, the MES will receive common channel signals from each satellite. The unit will select the strongest signal unless told, by the operator, to select a

A Performance Verification Test consists of 3 main tests:

- CES-to-MES message
- MES-to-CES message (same message returned)
- Distress Alert Test

The entire test will take some minutes, depending on NCS/CES traffic load and link performance. The procedure is outlined below.

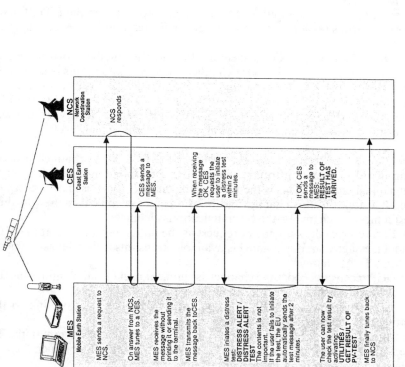

Fig. 8.5 A performance verification test. (Courtesy ABB NERA AS)

specific satellite. As the mobile moves to the outer edges of satellite coverage, signal level falls and the MES looks for a new satellite.

Once the equipment has been logged-in, it is the responsibility of the operator to select a LES, from the many which may be available if a communications link is required.

Logging-out

If the equipment is to be switched off because it is not required, it is essential that the system be logged-out. Logging-out causes the MES DTE to transmit a log-out signal to the NCS. This act keeps the NCS, and Inmarsat, informed of the operational status of the MES and enables traffic to be re-routed.

If no logging-out of the MES takes place, all the elements in the Inmarsat-C system still register the terminal as available for traffic. The NCS will continue to call the MES and instruct it to tune to a TDM channel, for traffic, causing unnecessary signalling. Because the MES does not send an 'assignment response' message, the LES assumes that the to-mobile announcement did not get through. After a few minutes the LES repeats the sequence, sending another MES 'status request' call to the NCS and awaits an 'assignment response' from the MES. This signalling may continue for up to an hour until eventually the LES stops attempting to communicate with the MES.

This unnecessary signalling uses satellite transponder time and consequently the MES may be charged even though the message is never received.

Call set-up procedures

From-mobile call
The message to be sent is input to the equipment and stored in the DTE, for store-and-forward transmission, until the system indicates that it is ready to receive the traffic.

The DCE receiver checks its memory to find the number corresponding to the frequency of the selected LES TDM channel. The DCE knows this because it has been continuously monitoring the NCS common signalling channel since logging-in.

The DCE now changes its tuning from the NCS common channel to the LES TDM channel. (Depending upon the make of terminal, the operator may see this act on the VDU display as the channel number changes and the signal levels vary at change-over.)

When the DCE receiver is synchronized to the LES TDM channel, it decodes the bulletin board information to find the number corresponding to the frequency of the SES signalling channel. This is the LES calling channel to be used.

The DCE tunes to this channel and transmits an assignment request to the LES to request a message channel—the communications channel.

The LES receives the assignment request as the initial indication that the MES wants to send a message and, if a channel is available, transmits a channel assignment packet in the LES TDM channel.

The DCE receiver decodes the channel assignment packet and identifies the frequency of the channel to be used for transmission.

The LES now sends an MES busy packet over the interstation link to the NCS, indicating that the MES is busy.

The NCS uses the MES busy packet to update its database concerning this MES.

The MES DTE now passes the message to the DCE transmitter which then sends the message in data packets on the message channel to the LES.

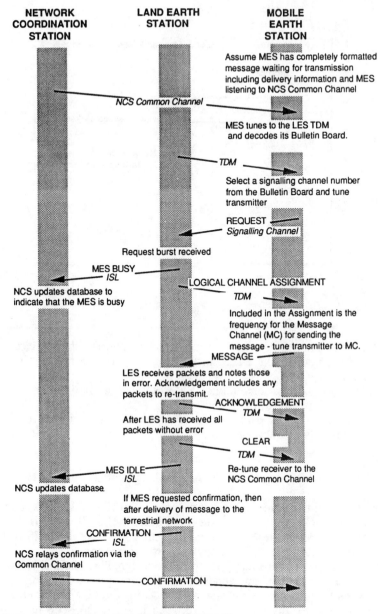

NETWORK COORDINATION STATION

LAND EARTH STATION

MOBILE EARTH STATION

Assume MES has completely formatted message waiting for transmission including delivery information and MES listening to NCS Common Channel

NCS Common Channel

MES tunes to the LES TDM and decodes its Bulletin Board.

TDM

Select a signalling channel number from the Bulletin Board and tune transmitter

REQUEST
Signalling Channel

Request burst received

MES BUSY
ISL

NCS updates database to indicate that the MES is busy

LOGICAL CHANNEL ASSIGNMENT
TDM

Included in the Assignment is the frequency for the Message Channel (MC) for sending the message - tune transmitter to MC.

MESSAGE

LES receives packets and notes those in error. Acknowledgement includes any packets to re-transmit.

ACKNOWLEDGEMENT
TDM

After LES has received all packets without error

CLEAR
TDM

MES IDLE
ISL

Re-tune receiver to the NCS Common Channel

NCS updates database

If MES requested confirmation, then after delivery of message to the terrestrial network

CONFIRMATION
ISL

NCS relays confirmation via the Common Channel

CONFIRMATION

Fig. 8.6 From-mobile call procedure. (Courtesy Inmarsat)

The LES receives, decodes and stores the incoming data packets, and performs error checking and correction processes. If any errors are detected the following occurs.

- The LES returns an acknowledgement packet to the SES with details of the data packets received in error.
- The DCE receiver decodes this information and instructs the DCE transmitter to re-transmit the packets received in error.
- This sequence may be repeated.

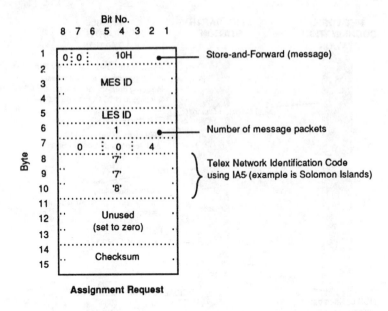

Fig. 8.7   Message transfer frame. (Courtesy Inmarsat)

When the LES is satisfied that the message is correct, it returns a channel clear packet on the LES TDM channel to the MES, instructing it to retune to the NCS common signalling channel.

The channel clear packet also contains details of the confirmation of reception of the message which is printed for information.

The LES sends an MES idle packet to the NCS announcing that the MES is now free.

The NCS updates its database accordingly.

To-mobile call

The idle MES is tuned to the NCS common signalling channel and updates its memory accordingly.

The message, from the terrestrial network, appears at the LES which, upon receipt, uses the ISL to check the status of the designated MES. If the MES is free, an announcement is made on the NCS common channel advising the MES of the traffic and telling it to tune to the LES TDM channel for information.

Once the LES bulletin board has been read and a signalling channel has been found the MES tunes to that channel and sends an assignment response message to the LES. The LES now sends the traffic, as message packets, using the TDM channel and requests an acknowledgement. If any packets are received in error the MES tells the LES to re-transmit those packages. If no repeat is required an acknowledgement is sent and the LES responds by sending a clear message to the MES on the TDM channel.

The MES now re-tunes to the NCS common signalling channel. The LES advises the NCS that the MES is idle and the NCS updates its database accordingly.

Data reporting and polling

This Inmarsat-C service is provided to enable land-based companies or government agencies to receive data directly from mobiles. As an example, shipping companies may

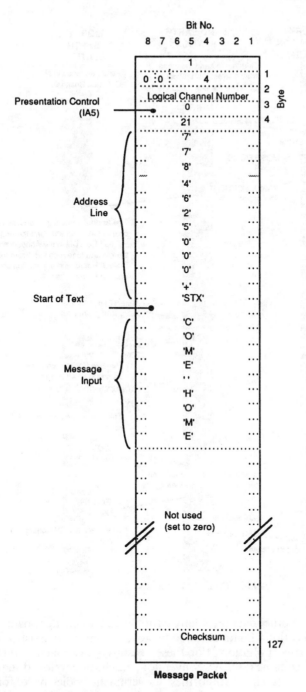

Fig. 8.7 Message transfer frame. (Courtesy Inmarsat)

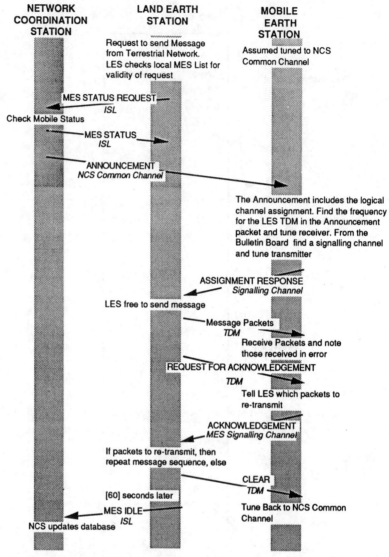

Fig. 8.8   To-mobile call procedure. (Courtesy Inmarsat)

require positional, or other data on a daily basis, leading to greater operating efficiency. An important application of data required by governments is to assist ships in distress, and to prevent marine pollution. For these reasons, governments, through various international organizations (SOLAS convention, etc) have developed requirements for ships to make reports to shore authorities. Government bodies have, for some years, been involved in receiving information from ships primarily for weather reporting. Position reporting is not new. International ship position reporting data has, in the past, been sent by Morse code and used by the US Coast Guard service as part of their AMVER organization.

Ship data may be interfaced directly with an Inmarsat-C terminal from the various navigational, engine or weather sensors carried on board. There are two ways in which these data may be sent: either automatically or manually at pre-determined times from the MES, or alternatively, the MES may be interrogated (polled) from shore. When a polling command is sent to an MES the unit will respond by automatically returning data in a pre-determined format.

The poll command contains information about how and when the MES should respond. The types of polling available are as follows.

- Individual polling—An explicit poll command to one MES. If the MES is busy the poll is queued until the terminal is idle.
- Group polling—A single poll command to a group of mobiles. Typically a fleet call.
- Area polling—A single poll to a number of mobiles located in a specific geographical area.
- Enhanced group calling (EGC)—This is a message broadcast service to Inmarsat-C users from land-based information services.

There are two types of EGC calls:
- FleetNET, which is for commercial use by individual companies to their mobiles; and
- SafetyNET, which is provided exclusively for maritime users. See Chapter 1.

It is possible that future polling commands may include remote monitoring and control instructions whereby data polled from an SES is used by an office ashore for remote control of some of the operating parameters of a vessel.

Data reporting and polling formats use standardized short binary encoded data messages for the transfer of information. Data reports are limited to three data packets which, with header information, provide 32 bytes of data. In addition, macro-encoded messages (MEMs) may be used. These are pre-defined text messages which are compressed to provide more efficient use of reporting formats. In the Inmarsat standard maritime report, a total of 128 MEM codes are available, of which about half are defined by Inmarsat.

Mobile earth stations

Inmarsat-C MESs are classed as follows.

- A class 0 MES: EGC receivers may be fitted in the following ways.
  (i) Option 1: as a stand-alone EGC receiver, without communications options; or,
  (ii) Option 2: as a stand-alone EGC receiver added to an Inmarsat-A terminal.
- A class 1 MES is used for two-way communications to/from the mobile but is not able to receive EGC messages.
- A class 2 MES is capable of two modes of operation.
  (i) As in class 1 and also for receiving EGC messages.
  (ii) Exclusively for EGC reception.
- A class 3 MES has two independent receivers, one for EGC reception and one for communications, thus making simultaneous operation in both modes possible.

There are two parts to an Inmarsat-C SES. The data terminal equipment (DTE) which forms the below decks electronics unit and the data circuit terminating equipment (DCE) which is the antenna, transmitter and receiver.

The DCE part of an Inmarsat-C SES normally comprises a right-hand circular (RHC) polarized omnidirectional antenna and electronics fitted inside a protective radome.

In practice, the antenna should be mounted in such a position that it possesses an

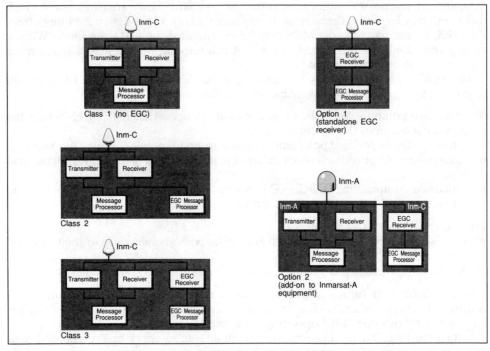

Fig. 8.9   Different classes of Inmarsat-C ship earth station

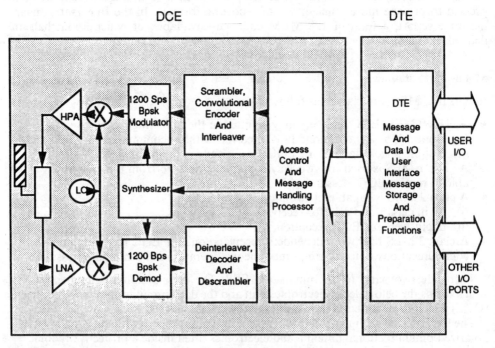

Fig. 8.10   Inmarsat-C mobile earth station example diagram

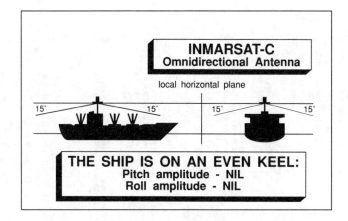

Fig. 8.11 Inmarsat-C omnidirectional antenna location

unrestricted view of the sky from −15° elevation, and through 360° in azimuth. There should be no metal objects within 3 m of the antenna which would cause blind arcs.

Downlink signals from the antenna are coupled, via a low-noise amplifier, to the mixer where they are converted to an intermediate frequency. BPSK demodulation produces the raw data which is descrambled before being passed down to the DTE for processing. The DTE comprises a laptop computer with NBDP and possibly other peripherals attached. Decoded messages are either stored, with a message displayed to warn the operator, or are printed directly upon receipt.

Messages to be sent are input via the operator keyboard and prepared for transmssion using word processing, after which they are stored. Using the Inmarsat-C 'store-and-forward' system the MES automatically gains access to the satellite at a convenient time and sends the message to a selected LES for forwarding.

A number of Inmarsat-C SES installations provide full integration with a global position system (GPS) receiver to give instant navigation position fixing

SES VDU display

The video display unit (VDU) provides the operator with an extremely versatile information display. The Saturn-C uses a laptop PC computer which provides full editing facilities, communications control and information display.

The Saturn-C screen display carries all the vital information for system operation. The top menu line has five pull-down menus which may be demanded by the operator. The communications message display (information window) is shown blank. The illustrations in figure 8.12 demonstrate how versatile a computer screen is when used in this context. Many readers will be computer literate and consequently will feel happy interpreting the displayed information.

As an example of the use of pull-down menus, part of the system configuration display is shown in figure 8.13.

The ship is currently logged into the AORW. By using one or more of the commands shown it is possible to change ocean regions and NCSs, providing of course that the ship is within the footprint of another satellite ocean region. Figure 8.14 shows part of the message display which announces one of 43 different system messages requiring action from the operator.

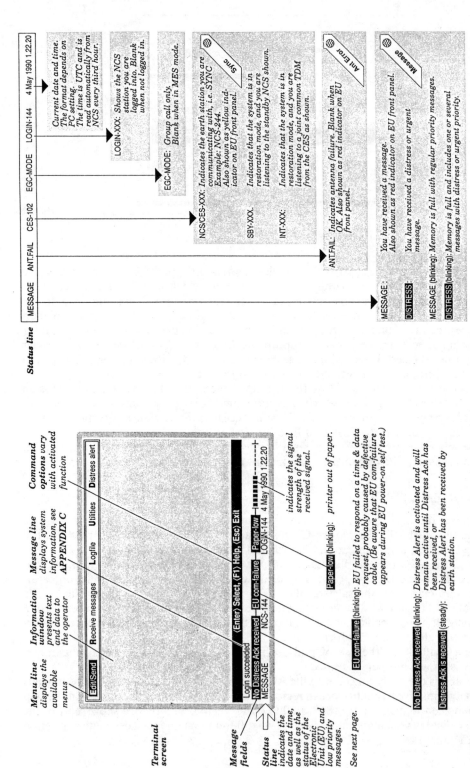

Status line

| MESSAGE | ANT.FAIL | CES-102 | EGC-MODE | LOGIN-144 | 4 May 1990 1.22.20 |

Current date and time. The format depends on PC setting. The time is UTC and is read automatically from NCS every third hour.

LOGIN-XXX: *Shows the NCS station you are logged into. Blank when not logged in.*

EGC-MODE: *Group call only. Blank when in MES mode.*

NCS/CES-XXX: *Indicates the earth station you are communicating with, i.e. SYNC Example: NCS-244. Also shown as yellow indicator on EU front panel.*

SBY-XXX: *Indicates that the system is in restoration mode, and you are listening to the standby NCS shown.*

INT-XXX: *Indicates that the system is in restoration mode, and you are listening to a joint common TDM from the CES as shown.*

ANT.FAIL: *Indicates antenna failure. Blank when OK. Also shown as red indicator on EU front panel.*

MESSAGE: *You have received a message. Also shown as red indicator on EU front panel.*

DISTRESS: *You have received a distress or urgent message.*

MESSAGE (blinking): *Memory is full with regular priority messages.*

DISTRESS (blinking): *Memory is full and includes one or several messages with distress or urgent priority.*

Terminal screen

Menu line displays the available menus

Information window presents text and data to the operator

Message line displays system information, see APPENDIX C

Command options vary with activated function

Edit/Send   Receive messages   Logfile   Utilities   Distress alert

(Enter) Select, (F1) Help, (Esc) Exit

Login succeeded

No Distress Ack received — EU com-failure — Paper-low

MESSAGE   NCS-144   LOGIN-144   4 May 1990 1.22.20

indicates the signal strength of the received signal.

Paper-low (blinking): *printer out of paper.*

EU com-failure (blinking): *EU failed to respond on a time & data request, probably caused by defective cable. (Be aware that EU com-failure appears during EU power-on self test.)*

No Distress Ack received (blinking): *Distress Alert is activated and will remain active until Distress Ack has been received, or*

Distress Ack is received (steady): *Distress Alert has been received by earth station.*

Message fields
See next page.

Status line *indicates the date and time, as well as the status of the Electronic Unit (EU) and low priority messages.*

Fig. 8.12 Saturn-C screen display and status line explanation. (Courtesy ABB NERA AS)

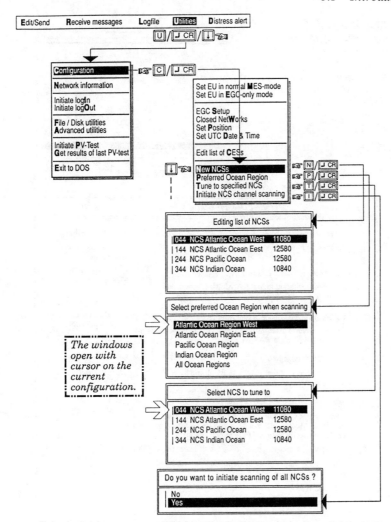

Fig. 8.13 Saturn-C display showing part of the configuration mode in which NCSs and ocean regions are selected. (Courtesy ABB NERA AS)

*The messages 1-15 listed below appear **in the message line** on the screen display.*

The messages contain system information and are displayed for a period of 1 minute.

Ref	Message/Description	Action
1	New message from shore has arrived A message has arrived.	Retrieve message: *Receive messages / Read new message.*
2	Distress test is initiated automatically The operator failed to manually initiate a distress test within 2 minutes during a distress test sequence.	Do nothing, wait until PVT is finished.
3	Position not updated, done by NMEA device Position update attempt failed because EU is set to accept position from an NMEA device only.	If NMEA device is not connected, enable manual setting: Utility /Configuration/Position.
4	Result of test has arrived Denotes arrival of test results to EU from NCS/CES during a PV-test.	Read the results: *Utilities / Get results of last PV-test*
5	Aborting transmission succeded: [...reason...] Denotes that CES has accepted the abort call from Saturn C. "Abort Call" has been initiated by the operator or automatically from EU (due to protocol error).	
6	Sending the message succeeded, message ref. number is: XXXXX CES reports the message as completely received and also reports the message reference number	
7	Login succeeded NCS confirms login after such request	

Fig. 8.14   Saturn-C display showing the first seven of over 40 systems messages

9
The Inmarsat-M system

9.1 Introduction

It must be said immediately that, although it contains a distress function, this system does not meet the rigid criteria set by the IMO for operation within the GMDSS. This factor is particularly important when considering a new maritime installation. However, the system has been included in this book because it is new, it exists and it is possible that its parameters will, at some time in the future, satisfy the IMO criteria for GMDSS satellite communications.

Inmarsat-M offers a mobile satellite communications system of the highest quality whilst keeping costs to a minimum. The system became operational in parallel with the Inmarsat-B system, and is intended to provide low-cost, lightweight MESs offering good quality voice and low-speed data communication services for users who require voice-grade communications interfaced with the PSTN from a compact MES.

Two types of MES are available with the system:

- a maritime mobile MES, for use on small vessels where lighter and more compact ADE is needed than that provided by the Inmarsat-A and B systems; and,
- a land mobile MES, which may be vehicular mounted or transportable.

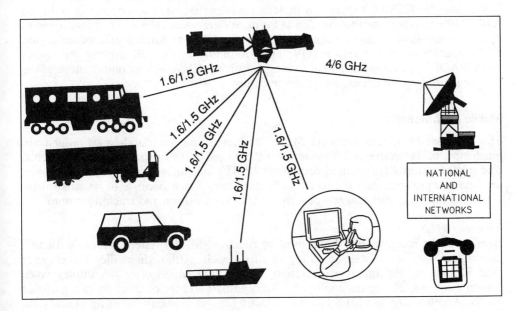

Fig. 9.1 Inmarsat-M network configuration

The compactness of Inmarsat-M MES equipment, it is anticipated, will lead to a rapid expansion in the land mobile market—an area which has previously been difficult to develop, for voice communications, using the large mobile units of the Inmarsat-A system. Indeed, Inmarsat market research has predicted that there will be approximately 600,000 Inmarsat-M MESs in use by the year 2005. Clearly, if this forecast is anywhere near accurate the Inmarsat-M system will be a major provider of satellite mobile communications well into the next century.

Outline system description

Intentionally, the Inmarsat-M and B systems share much commonality, thus enabling design costs to be kept to a minimum. Inmarsat-M ground and space segments are the same as those used in the Inmarsat-B system, as is the access control and signalling subsystem. Digital speech processing is used which, although the coding rate is slower than that used in Inmarsat-B, at $6.4\,\text{kbits s}^{-1}$ (including error correction coding) produces speech as good as, or better than, most cellular mobile radio systems. Energy efficiency, very important for an MES operating from batteries, and bandwidth efficiency, are both greatly improved over the ageing Inmarsat-A system by the use of modern digital technology.

The space segment

This is identical to that used in the Inmarsat-B system.

The ground segment

Inmarsat-M uses the same ground operational organization of LES, NCS, OCC, SCC and TT&C stations, although the designated NCS for each ocean region may be different from the other systems. Inmarsat-M became a fully global system in Autumn 1993 when the CES at Yamaguchi in Japan was inaugurated to operate in both the IOR/POR ocean regions and the CES at Perth, Western Australia was commissioned to operate in the IOR. These stations along with the CESs at Santa Paula, ocean region POR, Southbury, ocean region IOR, and Goonhilly in the UK serving the region AORE/AORW provide full global coverage. Further CESs serving one or more of the four ocean regions will come on stream in future.

Mobile earth stations

The Inmarsat-M system supports MESs, both maritime and land-based, which are much more cost effective and compact than those used in the Inmarsat-A and B systems. The smaller externally mounted equipment (EME) unit includes a steerable parabolic antenna of approximately 0.3 to 0.5 m in diameter, which because of its small mass requires a much smaller and less complex satellite acquisition and tracking system.

Maritime MESs
Because of the much smaller EME used by Inmarsat-M, the system is more attractive to companies or individuals operating smaller vessels. Although smaller and cheaper than SESs using the Inmarsat-B system, Inmarsat-M MESs offer full duplex voice communications, of a lesser quality, but sufficiently good for connection to a national PSTN. Additionally, the MES has the capacity for communications using digital data services at $2.4\,\text{kbits s}^{-1}$ which is slower than the Inmarsat-B system but is indeed

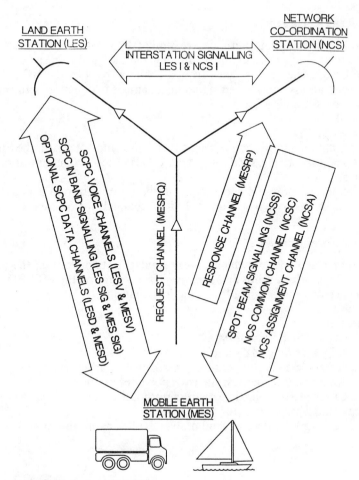

Fig. 9.2 Inmarsat-M channel configuration

faster than the Inmarsat-C system. Whilst the maritime MES is provided with a distress priority system, the reduced power and operating parameters of the MES leads to the slight possibility of a link being unavailable at high latitudes – low satellite elevation angles. Consequently, the Inmarsat-M maritime MES does not comply with the IMO requirements for distress and safety communications within the GMDSS radionet.

Land mobile MES
Land mobile MESs are still in the early stages of development. The EME design introduces a new set of design challenges. In most cases the MES is likely to be identical to the maritime MES. However, satellite acquisition and tracking techniques from a rapidly moving vehicle will require electro-mechanical systems which respond more rapidly to changes in azimuth and elevation than a maritime antenna. This antenna may be a parabolic type but, in the future, it is more likely to be a low mass phased array or a crossed dipole array. The transmit signal beamwidth is wide in elevation and narrow in azimuth in order to maintain satellite lock and simplify many of the tracking requirements.

Access control and signalling

Access control and signalling methods are similar to those used in the Inmarsat-B system in that communications channels are demand assigned in response to requests from a MES. Frequencies are assigned from a common pool by the NCS. The system uses the following access control and signalling channels in common with Inmarsat-B:
NCSC, NCSA, NCSI, and NCSS
LESA and LESI
MESRQ and MESRP.
SCPC (voice) communication channels are titled **LESV** and **MESV**, in the forward and reverse directions respectively. Optional TDMA/TDM (data) communication channels are labelled **LESD** and **MESD**.

There are additional optional network co-ordination registration acknowledgement (NCRA) and ocean region registration (MESRR) channels for future use.

Call set-up procedures from the MES

As with the Inmarsat-B system which is very similar in its operation, the sequence is initiated by the MES which sends an *access-request message* on the MESRQ channel to the LES. The message contains the following information:
LES identification
MES identification
Request service type (telephony or data)
Priority (for use by maritime MES only)
MES antenna AZ angle zone (for frequency assignment purposes)
MES antenna AZ angle (for maritime distress purposes only)
MES antenna EL angle (for maritime distress purposes only)
Satellite spot beam identification (for future use)
Received signal strength report (for land mobile MES forward link power control

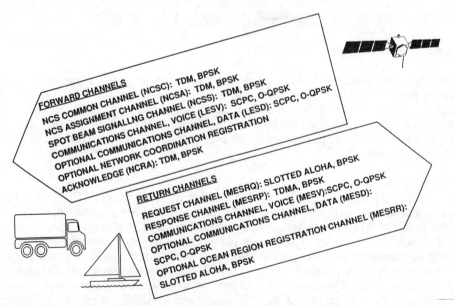

Fig. 9.3　Inmarsat-M channel description, mobile satellite links

The process then continues as described in the Inmarsat-B system. Figure 9.3 shows the full forward and return channels along with the access/modulation methods used in the system.

Spot beam signalling and network coordination registration channels will be used when the third generation of Inmarsat satellites, INM3, is in operation.

9.2 Which Inmarsat system?

Once the decision to fit a satellite communications terminal has been made, the next decision is which Inmarsat system is best suited to individual requirements.

Essentially, the criteria for such a decision should be based on sound economic principles whilst taking into account the need for operation within the GMDSS. Not only the cost of installation but also the cost of daily use should be considered.

The purchase cost of an SES varies widely depending upon which manufacturer, the services provided and the system chosen. Operating costs vary considerably depending upon which service is used, which system is fitted and the chargeable duration of the communications link. The average daily use of an SES is about 8 minutes of telephone, fax or data and 6 minutes of telex. Obviously there are 'high traffic' ships which use their SES far in excess of these figures and other ships which will hardly ever use the service.

The flow charts which follow are based not on the initial cost of equipment or that of a specific manufacturer. The criteria used is essentially the extent to which an SES will be used, the services which are required and the question of GMDSS compliance. Services required may be voice, data, fax and/or telex. Additionally, it is assumed that the new installation will be made on or after January 1994 when approved Inmarsat-B

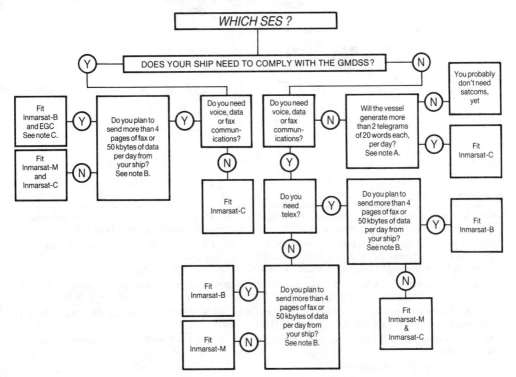

Fig. 9.4   Which ship earth station?

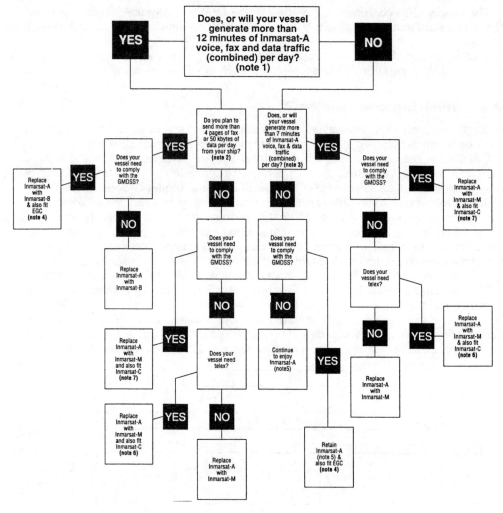

Fig. 9.5 Have you been planning to replace your Inmarsat-A? (Courtesy Inmarsat)

Notes to figure 9.5

Note 1: A vessel generating a combined total in excess of 12 minutes of Inmarsat-A voice, fax and data traffic per day would, because of its lower charges, find it economical to change to Inmarsat-B.

Note 2: A low traffic vessel producing less than four pages of fax or equivalent data per day with no prospect of any increase, would find no difference between Inmarsat-M and Inmarsat-B. However for greater amounts of traffic move to the higher transmission speed of Inmarsat-B.

Note 3: If the vessel generates a combined total of between seven and twelve minutes of Inmarsat-A voice fax and data traffic per day, Inmarsat-M would be satisfactory – if not fitting to GMDSS requirements.

Note 4: If the vessel is subject to GMDSS and will sail in areas where the Navtex service is not provided, a full Inmarsat-C or a stand-alone EGC receiver MUST be fitted

Note 5: As you are a low traffic ship not requiring to comply with the GMDSS any Inmarsat standard would suffice.

Note 6: Inmarsat-C supports telex.

Note 7: Inmarsat-C meets carriage requirements for the GMDSS.

and M SESs will be available. Because the new systems will eventually replace Inmarsat-A it is anticipated that brand new Inmarsat-A SESs will not be fitted after that date.

● **Note A.** A vessel transmitting more than two telegrams, each of more than 20 words, per day should recover the capital outlay of an Inmarsat-C SES in approximately one year when compared with the cost of using terrestrial radio-telegraphy. This fact does not include the large on-going costs of carrying a dedicated radio officer to operate telegraphy equipment.

● **Note B.** Inmarsat-M's data and fax speeds of up to 2.4 kbits s^{-1} will satisfy the requirements of the majority of Inmarsat users. High-volume data users should fit Inmarsat-B.

● **Note C.** Ships operating within the GMDSS radionet must be able to receive maritime safety information (MSI). A dedicated enhanced group calling (EGC) receiver or an Inmarsat-C SES will receive this service, which is transmitted through the international SafteyNET service.

Figure 9.5 provides possible solutions for replacing an ageing Inmarsat-A SES. The main criteria for choice are as before: compliance with the GMDSS, extent of use and services provided. Figures 9.4 and 9.5 should in no way be considered to be the definitive solution to fitting requirements. Specific communications requirements will require custom devised installations. Check with your supplier but do consider your future expanded communications requirements before making a decision.

10
Satellite mobile frequency bands

In order for any communications system to expand within the technology available, to keep pace with ever increasing demands for channels, it is critically important that more frequency spectrum is made accessible. The World Administrative Radio Conference met in 1992 (WARC92) at Torremolinos in Spain to consider the problems encountered by mobile satellite services (MSS). As a result the frequency bands available for MSS have been increased.

The original frequency bands allocated to mobile satellite communications were 1535.000 – 1542.500 MHz (receive) and 1636.500 – 1644.500 MHz (transmit). A small band of frequencies indeed in which to offer a service on a global scale. Because of the increase in the number of SESs in use, and other systems being introduced, the band available has been increased by previous meetings of WARC to 63 MHz, 1530.000 – 1548.000 MHz (receive) and 1626.500 – 1649.500 MHz (transmit).

After WARC92 a further 33 MHz of global frequency spectrum has immediately become available for use by the MSS, thus increasing the spectrum available by over 50%. The bands, to be used by the new Inmarsat INM3 satellites, are 1524.000 –

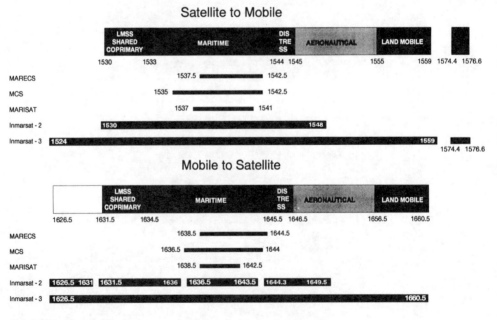

Fig. 10.1 The Inmarsat space segment channel frequencies

1559.000 MHz receive and 1626.500 – 1660.500 MHz transmit. Further huge increases of 80 MHz near 2 GHz, and 70 MHz near 2.5 GHz, will become accessible after the year 2005, ensuring that MSS will expand to meet future needs, as new technology becomes available.

11
The future

The future direction being taken by Inmarsat is based on the recommendations of a development group—entitled Project 21—which has the task of planning and implementing the systems required for a global pocket-sized satellite telephone service by the end of the decade. Miniaturization is a constant fact of electronic engineering. Over the past decade we have seen satellite and terrestrial communications equipment shrink by approximately 50%. The level to which miniaturization can be taken is, of course, limited. The equipment is ultimately going to be used by human beings and consequently must be ergonomically designed for ease of operation. Additionally, the enclosed battery tends to be bulky and accounts for almost one-third of the bulk of a hand-held unit.

Advances in space technology have now enabled much larger satellites with higher powered transmitters to be launched as witnessed by the inauguration of the new Inmarsat INM2 spacecraft. Much greater capacity Inmarsat INM3 satellites are being developed for launch in 1994. INM3 satellites will also carry a navigation package enabling communications and navigation services to be available from the same satellite.

The establishment of efficient higher powered satellite transponders means that much smaller earth-bound transceivers using small omni-directional antennae becomes a reality. Inmarsat's space segment of the future will not rely solely on the use of geostationary satellites. Many proposed future services can be better served using a combination of geostationary (GEO) satellites, high elliptical orbiting (HEO) satellites and low earth orbiting (LEO) satellites. The hand-held voice units of the next decade will probably use a combination of GEO and LEO satellites in order to provide good coverage and improve frequency spectrum efficiency.

Inmarsat's expansion over the next decades will also include providing services to briefcase phones, global paging, navigation, position reporting, voice broadcasts and a continuing involvement in the GMDSS for worldwide distress alerting and calling.

Section Three

TERRESTRIAL COMMUNICATIONS

12
Signal propagation and the radio spectrum

12.1 Introduction

This section outlines the basic principles of communications, the communications systems and the services used in the terrestrial segment of the GMDSS radionet.

The use of radiowaves for terrestrial international communications causes major problems, particularly in the area of frequency allocation and interference. However, in many cases it is not the technical problems which cause difficulty with terrestrial international radiocommunications, but the political problems. The international governing body for radio communications services is the International Telecommunications Union (ITU) which, quite rightly, strictly regulates the use of the limited resource which we call the radio frequency spectrum. It is by strict control of the use of this spectrum that all national and international radiocommunications are able to take place without severe interference occurring. Radiowaves cannot and do not respect international boundaries and, consequently, disputes arise between nations over the use of radio frequencies. These disputes are settled by the ITU through various committees and affiliated organizations. All users of radiocommunications systems must be aware that they are licensed to use only specific frequencies and systems in order to achieve information transfer. It would be chaos if this were not so. Essential services, aeronautical, maritime or land based, would not be able to operate otherwise and lives could well be put at risk. However, despite the ongoing political and technical problems it is possible to achieve good quality truly global communications using the HF bands and the earth's ionospheric layers.

12.2 Maritime communication systems and their frequencies

Maritime radiocommunications requirements have always posed unique problems for the shipboard operator. A ship at sea presents numerous difficulties to the radiocommunications design engineer. The ship is constructed of steel which, when floating in salt water, becomes a very effective electromagnetic screen capable of rejecting or reflecting radiowaves. Additionally, modern ocean-going vessels are streamlined and designed to enable the efficient use of fuel. This, in turn, means that the traditional sturdy structures for holding antenna systems—funnels or masts—have been reduced or removed. Consequently, shipboard antenna systems tend to be less than efficient, giving rise to difficulties in both transmission and reception.

The systems used within the GMDSS radionet are detailed in Chapter 1. Listed below is a summary of those terrestrial services:

- Navtex data on the medium frequency 518 kHz,
- voice, radiotelex and digital selective calling (DSC) using the medium frequency band 1.6 MHz to 3.4 MHz,

- voice, radiotelex and DSC using the high frequency bands between 3 MHz and 30 MHz
- voice and DSC in the very high frequency band 30 MHz to 300 MHz,
- RADAR SART on the frequency of 9 GHz.

It should be noted that the combined systems needed for operation within the GMDSS require the use of a very wide range of frequencies from the relatively low frequency 518 kHz MF up to the L-band frequency of 9 GHz.

In each case, the carrier frequency used has been chosen to satisfy two main criteria, those of geographical range and the ability to carry the relevant information. The geographical range of a radiowave is affected by many parameters, but to provide an understanding of the GMDSS systems operation, range may basically be related to the choice of frequency band, which in turn determines the method of radiowave propagation.

12.3   Radiowave radiation

The propagation of radiowaves is a highly complex natural phenomenon. The methods utilized for the transmission of radiowaves and the natural parameters affecting the propagated radiowave will be simplified to describe this phenomenon for the understanding of technical operators using the GMDSS radionet.

Energy is contained in a propagated radiowave in two forms, electrostatic energy and electromagnetic energy.

The radiation of energy from a simple antenna may be described by considering a centre-fed dipole antenna, which is shown electrically in figure 12.1(a). Antenna systems are described in detail later in this chapter.

The antenna shown in figure 12.1 is formed of two coils, each end of which is at the opposite potential to the other with reference to the centre point. As a complete unit, the antenna forms a tuned circuit which is critically resonant at the carrier frequency to be radiated. The two plates, one at each end of the coil assembly, form a capacitor. Radio frequency current, from the output stage of a suitable transmitter, shown here as a generator, is applied at the centre of the two coils. One of the basic electrical laws of

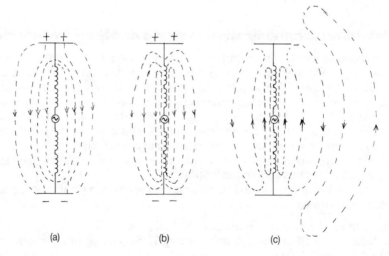

(a)   (b)   (c)

Fig. 12.1   Radiowave radiation from a centre-fed dipole antenna

physics states that whenever an electron has its velocity altered by an accelerating force there will be a detachment of energy. In the case of an antenna system this detachment is the energy which is lost from the transmitter and radiated as electrical energy into the atmosphere.

The diagrams in figure 12.1 show the distribution of the electric field around an antenna caused when an oscillatory radio frequency is applied to it. In figure 12.1(a) the top plate of the antenna is instantaneously driven positive with respect to the base plate and the current flow in the wire is zero. At this instant the field produced is entirely electric and, consequently, electrostatic lines of force are as shown in the diagram. After the peak of the signal has passed, electrons will begin to flow upwards to produce a current flow in the wire. The electric field will now start to collapse (figure 12.1(b)) and the ends of the lines of force come together to form loops of electrostatic energy. After the potential difference (positive top plate to negative base plate) across the two plates of the effective capacitor has fallen to zero, current continues to flow and, in so doing, starts to charge the capacitor plates in the opposite direction. This charge forms new lines of force in the reverse direction to the previous field, negative to the top plate and positive at the base. The collapse of the initial electrostatic field lags the change in potential which caused it to occur and, consequently, the new electric field starts to expand before the old field has completely disappeared. The electric fields thus created (figure 12.1(c)) will be caused to form loops of energy, with each new loop forcing the previous loop outwards away from the antenna. Thus, radio frequency energy is radiated as closed loops of electrostatic energy.

Because a minute current is flowing around each complete loop of energy, a magnetic field will be created around the loop at 90° to it. The magnetic lines of force produced around the vertical electric field created by a vertical antenna, will be horizontal. Two fields of energy in space quadrature have thus been created and will continue in their relative planes as the radiowave moves away from the transmitting antenna.

The electric and magnetic inductive fields are in both time and space quadrature and are, in fact, 90° out of phase with each other in time and at right angles to each other in space.

The electric field is of greatest importance to the understanding of radiowave propagation, the magnetic field only being present when current flows around the loop as the electric field changes.

Figure 12.2 illustrates the relative directions of the electric field, the magnetic field and the direction of propagation. The oscillating electric field is represented by the vertical vector OE, the magnetic field by OH and the direction of propagation by OD. Another electrical law of physics, Fleming's Right-Hand Rule, which is normally applied to the theory of electrical machines, applies equally to the direction of propagation of the radiowave.

At any instantaneous point along the sinusoidal wave of the electric field it is possible to measure a minute current flow in the loop of energy. The current will be increasing and decreasing as it follows the rate of change of amplitude of the sinusoidal frequency (carrier wave) of the radiowave. It is this instantaneous change of current which, when in contact with a receiving antenna, causes a current to flow at the receiver input and a minute signal voltage, called an electro-motive force (emf), to appear across the antenna input.

The transmitted signal may now be considered to be a succession of concentric loops of ever-increasing radius, each one a wavelength ahead of the next. The radiowaves would be similar in appearance to the waves caused on the surface of a pond when a rock is tossed into it. Similarly, the radiowaves would diminish in amplitude as they moved away from the transmitter. Each loop is moving away from the transmitting

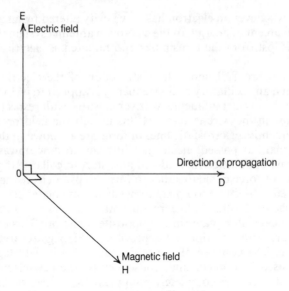

Fig. 12.2   The angular relationships of the E and H fields

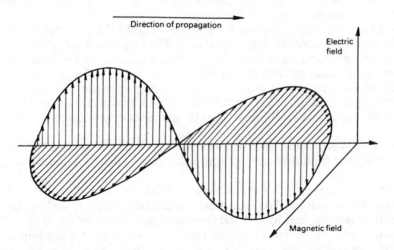

Fig. 12.3   Amplitude variations of the E and H fields

antenna at the speed of light in free space, which is usually approximated to be 300×10^6 m s^{-1}. It is common practice to call the leading edge of each loop a wavefront. The distance between each wavefront depends upon the frequency being radiated and is called the wavelength—termed λ (lambda). The wavelength as a parameter is dealt with later in this chapter when considering the design of GMDSS antenna systems.

Field strength/receiver sensitivity figure

A particularly important parameter, the anticipated radio signal field strength developed at the input to a receiver, called the power flux density (PFD), determines many of the receiver design parameters, in particular the sensitivity figure. In effect, the value of the PFD decreases as radio range increases but it is also subject to a number of other external effects.

Depending upon the amplitude of the power output of the transmitter, each propagated loop of energy will contain a finite energy level. Because the continually expanding wavefront covers an ever-increasing area as it travels away from the antenna, the energy is effectively spread over a greater area. Consequently, the signal induced in the receiving antenna will reduce as the distance between transmitter and receiver increases. Under free-space conditions, an inverse square law factor applies where the energy reduces with the square of the distance travelled. This assumes that all other factors affecting energy loss remain constant. In practice however, the nature of the medium through which the wave passes has a distinct effect on the amplitude of the signal received at the antenna. The intensity of an electromagnetic radiowave at any point in free space is generally expressed in terms of the strength of its corresponding electric field measured at right angles to the line of propagation. The units used for this measurement indicate the electro-motive force between two points in space one metre apart. As the electric field will be very small, whole units of voltage are not used. More typically, flux densities in microvolts will be stated.

The antenna circuit of a receiver is designed to detect the PFD of a specific radio signal at a given distance from a transmitter. In practice therefore, it is common for manufacturers to quote the minimum voltage which needs to be induced across the receiving antenna for a given field strength in order to produce a signal large enough to be detected by the demodulator in the receiver. The figure for input signal amplitude, termed receiver sensitivity, is typically within the range 1 to 10 μV, often standardized at 3 μV for a terrestrial communications receiver and very much smaller for a satellite system.

If a half-wave dipole antenna is placed into an electric field at the correct polarization angle of the received signal, the emf induced at its centre may be expressed as:

$$e = \frac{E\lambda}{\pi}$$

where e = emf at the dipole centre in whole units,
   λ = wavelength of the signal in metres,
   E = field strength in V m^{-1}.

In common with all signal carrying lines, the antenna must be connected to a matched feeder (coaxial cable) which will be correctly terminated at the input to the receiver. Thus, the input voltage available will be half of that received, $e/2$. By substituting frequency for wavelength and using fractions/multiples of whole units the equation now provides a more practical indication of microvolts at the receiver input.

$$V_r = \frac{47.8}{f} E$$

where V_r = microvolts of signal across the receiver input,
   f = frequency of received signal in MHz,
   E = field strength in μV m^{-1}.

It is often much easier in communications electronics to calculate in decibels. See Appendix 2.

Assuming that a base level of 1 μV m^{-1} is used for field strength and 1 μV for signal level, it becomes relatively simple to consider the various gains and losses of a receiving system for a given frequency using decibels. If each parameter is converted to dB the gains and losses in a system may be added and subtracted to produce the total.

$$V_r = V_o + V_i + G_r - L_a$$

all expressed as decibels
where in dB
V_r = receiver dB input relative to 1 μV,
V_o = dB input relative to 1 μV for 1 μV m^{-1} field strength derived from a chart,
V_i = actual incident field strength in dB relative to 1 μV,
G_r = gain in dB of the antenna relative to a half-wave dipole,
L_a = the feeder line attenuation.

12.4   Radio frequency spectrum

By referring to the table of the radio frequency spectrum it is possible to gain some initial idea of the approximate range over which radiowaves may be received. The table indicates how the frequency spectrum has been divided into bands which have well-known titles. For instance Search and Rescue (SAR) operations would most likely be conducted using VHF equipment. This equipment when used on channel 16 has a carrier frequency of 156.8 MHz which assigns it to the very high frequency (VHF) band. If all other parameters remain constant, the anticipated radio range of signals propagated on the VHF band, or those higher, is effectively by 'line-of-sight'. Consequently, SAR communications between a liferaft and a surface vessel could expect to have a range of about 10 to 12 nautical miles depending upon the system installation and the relative heights of the antennae. For radio ranges beyond the horizon to be achieved by earthbound stations it would be necessary to use repeater stations or utilize the LF, MF or HF bands, each of which has its own limiting characteristics. Mobile satellite communications tend to use the L band, whereas earth satellite stations use the C band for line-of-sight communications with a satellite.

As an example of the rigid control of the radio frequency spectrum, the maritime VHF band is carefully sub-divided into channels. There are ITU designated channels, American channels (which are similar to ITU channels), Canadian channels, weather channels and some private channels. ITU channels are in blocks 01 to 28 and 60 to 88. Some channels are allocated for duplex operation and some simplex. Some, like channel 70 are exclusively used for specific functions within the GMDSS radionet.

A similar but very much larger designation arrangement exists for the MF/HF bands. Channels are numbered within a band. As an example, channel number one in the 4 MHz band is 401 and in the 8 MHz band is 801 etc.

Spectrum management

Radiowaves do not respect international boundaries and, consequently, an international framework has been established in order to control the use of frequencies, the standards of manufacture and the operation of radio equipment in order to limit the likelihood of interference. The forum for reaching international agreements on the use of the radio frequency spectrum is the International Telecommunications Union (ITU). Membership of the ITU is dependent upon acceptance of the strict convention which exists to uphold the regulations laid down by the various conferences and meetings of the ITU and confirmed at Plenipotentiary conferences.

The radio spectrum management policies agreed among the signatories of the convention are published by the ITU as international radio regulations. The international 'Table of Frequency Allocations' thus produced provides the framework for, and the constraints on, national frequency use and planning. The Table of Frequency Allo-

THE RADIO FREQUENCY SPECTRUM		
Band Name and Frequency	Propagation Modes	Principal Characteristics
Very Low Frequency. 3 – 30 kHz	Large surface wave.	Very high power and large antennae needed.
Low Frequency. 30 – 300 kHz.	Surface wave. Some sky wave returns.	Limited number of channels. Subject to fading.
Medium Frequency. 0.3 – 3 MHz.	Surface wave during day. Some sky wave returns at night.	Long range at night. Subject to fading.
High Frequency. 3 – 30 MHz.	Sky waves returned over long distances.	Global ranges using ionospheric returns.
Very High Frequency. 30 – 300 MHz.	Mainly space wave. Line of sight.	Range depends upon antenna height.
Ultra High Frequency. 0.3 – 3 GHz.	Space wave only.	Line of sight. Satellite and fixed link.
Super High Frequency. 3 – 30 GHz.	Space wave only.	Line of sight. Radar and satellite.
Extreme High Frequency. 30 – 300 GHz.	Space wave only.	Not used for mobile communications.

Fig. 12.4(a)   The radio frequency spectrum

cations and the radio regulations documents are revised at the World Administrative Radio Conferences (WARC) held at periods of five to ten years.

The administrative structure established by the ITU convention comprises; a Secretariat headed by the Secretary General, an Administrative Council, a registration board for radio frequencies and the consultative committees for radio and telecommunications.

The International Radio Consultative Committee (CCIR) forms study groups to consider and report on the operational and technical issues relating to the use of radio communications. The International Telecommunications Consultative Committee (CCIT) offers the same service for telecommunications. The study groups produce recommendations on all aspects of radio communications. These recommendations are considered by the Plenary Assembly of the CCIR and, if accepted, become incorporated into the radio regulations. Another sub group of the ITU , the International Frequency Registration Board (IFRB), considers operating frequencies, transmitter sites and the

ITU VHF CHANNEL ASSIGNMENT TABLE					
CH	TX(MHz)	RX(MHz)	CH	TX(MHz)	RX(MHz)
01	156.050	160.050	60	156.025	160.625
02	156.100	160.700	61	156.075	160.675
03	156.150	160.750	62	156.125	160.725
04	156.200	160.750	63	156.175	160.755
05	156.250	160.800	64	156.225	160.825
06	156.300	156.300	65	156.275	160.875
07	156.350	160.950	66	156.325	160.925
08	156.400	156.400	67	156.375	156.375
09	156.450	156.450	68	156.425	156.425
10	156.500	156.500	69	156.475	156.475
11	156.550	156.550	70	156.525	156.525
12	156.600	156.600	71	156.575	156.575
13	156.650	156.650	72	156.625	156.625
14	156.700	156.700	73	156.675	156.675
15	156.750	156.750	74	156.725	156.725
16	156.800	156.800	75	—	—
17	156.850	156.850	76	156.825	156.825
18	156.900	161.500	77	156.875	156.875
19	156.950	161.550	78	156.925	161.525
20	157.000	161.600	79	156.975	161.575
21	157.050	161.650	80	157.025	161.625
22	157.100	161.700	81	157.075	161.675
23	157.150	161.750	82	157.125	161.725
24	157.200	161.800	83	157.175	161.775
25	157.250	161.850	84	157.225	161.825
26	157.300	161.900	85	157.275	161.875
27	157.350	161.950	86	157.325	161.925
28	157.400	162.000	87	157.375	161.975
			88	157.425	162.025

CH. 70 is exclusively assigned for DSC.
CH.76 is exclusively assigned for NBDP
CH. 75 is a guardband for ch. 16.

Fig. 12.4(b) ITU VHF channel assignment table

location of satellites in orbit. Within Europe, a further body, the Conference of European Telecommunications Administrations (CEPT), assists with the implementation of the ITU radio regulations on a national level. Each country within Europe also possesses an organization established by Government to enact the radio regulations thus laid down. In Great Britain this becomes the responsibility of the Radiocommunications Division of the Department of Trade. In the USA, civil use of the radio frequency spectrum is controlled by the Federal Communications Commission.

Maritime mobile bands and frequencies

The frequency bands used for terrestrial maritime mobile communications have been fought for at WARC committee meetings by various national authorities. Stiff competition exists between the multitude of frequency user groups seeking increased access to the frequency spectrum.

Essentially, terrestrial maritime mobile users are allocated bands of frequencies throughout the spectrum according to a proven user need. Frequencies for terrestrial maritime communications are allocated in the bands, MF, HF and VHF.

- **The MF band 435 kHz to 526.5 kHz.** Traditionally used for Morse code communications but no longer applicable within the GMDSS. The spot frequency 518 kHz continues to be used for NBDP broadcast to ships. NAVTEX.
- **The MF band 1.6065 MHz to 3.8 MHz.** In fact the frequencies above 3.0 MHz should be labelled HF but the band is known as the MF band. Communications channels are allocated with a 3 kHz channel bandwidth, for radio-telephone and telex operations either in simplex or duplex mode for medium-range working. Hundreds of channels are used in a multiplexed frequency arrangement in order that adjacent stations do not interfere with each other's working. GMDSS spot frequencies within this band include; 2174.5 kHz for distress and safety traffic by NBDP; 2182.0 kHz for distress and safety traffic by radio-telephone; and 2187.5 kHz for distress and safety calling using DSC.
- **The HF band 3.155 MHz to 27.5 MHz.** The band is subdivided into the traditionally known, 4 MHz, 6 MHz, 8 MHz, 12 MHz, 16 MHz, 22 MHz and 25 MHz bands. Once again only small sections of each band are allocated to maritime mobile users. Communications channels are allocated with a 3 kHz bandwidth for radio-telephone, telex, facsimile and data global communications services using either simplex or duplex modes. Thousands of channels are available for global communications using sky wave propagation via the ionized layers. It is essential therefore that the allocated channels are strictly adhered to, otherwise total chaos will result. In practice, a system of interlocking on modern communications transceivers prevents out-of-band or cross-band working.

GMDSS spot frequencies within these bands include:
4207.5 kHz, 6312.0 kHz, 8414.5 kHz, 12,577.0 kHz and 16,804.4 kHz for distress and safety calls using DSC.
4210.0 kHz, 6314.0 kHz, 8416.5 kHz, 12,579.0 kHz, 16,806.5 kHz, 19,680.5 kHz, 22,376.0 kHz and 26,100.5 kHz for broadcast by shore stations of MSI by NBDP.
3023.0 kHz, 4125.0 kHz and 5680.0 kHz for SAR communications between maritime and aeronautical units.

- **The VHF band 156 MHz to 174 MHz.** Communications channels are allocated throughout the band at 12.5 kHz intervals. Each maritime channel is identified by a decimal number. The original numbering was 1 to 28. Subsequently, as a result of reducing the individual channel spacing from 25 kHz to 12.5 kHz, additional channels were made available interspersed between the existing ones. They are designated 60 to 88.

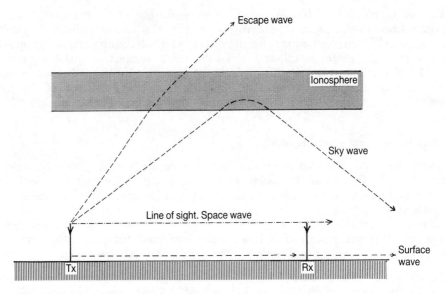

Fig. 12.5   Radiowave modes of propagation

GMDSS spot frequencies within the band include:
156.30 MHz (Channel 06) used for communication between ship and aircraft stations involved in SAR operations.
156.65 MHz (Channel 13) ship-to-ship communications relating to the safety of navigation.
156.80 MHz (Channel 16) International distress, urgency, safety and calling channel until 1999.
156.525 MHz (Channel 70) distress, urgency and safety calling channel using DSC.
   Additionally, the following GMDSS bands of frequencies are used.
406 MHz to 406.1 MHz; satellite EPIRBS earth-to-space direction.
1530 MHz to 1544 MHz; space-to-earth satellite communications.
1626.5 MHz to 1645.5 MHz; earth-to-space satellite communications.
9200 MHz to 9500 MHz; maritime surface radar including SAR operations and SART transponders.·

12.5   Radiowave propagation

When radiowaves are propagated from a transmitter antenna system they form one or more of three modes with one mode being dominant.
   The three modes of propagation are:

- surface wave propagation
- sky wave propagation, and
- space wave propagation.

   The surface wave is a radiowave which is modified by the nature of the terrain over which it travels. This can occasionally lead to difficulty for maritime communications where the wave travels from one medium to another, over a coastline for instance. The refraction caused in such a case leads to errors on a maritime radio direction finder.
   The space wave, when propagated into the troposphere by an earth surface station, is

subject to deflection by variations in the refractive index structure of the air through which it passes. This will cause the radiowave to follow the earth's curvature for a short distance beyond the horizon causing the radio horizon to be somewhat longer than the visible horizon. The effect will be known to navigators whereby the surface radar range of a maritime PPI radar picture extends slightly beyond the horizon. Space waves propagated upwards away from the troposphere may be termed free space waves and are primarily used for satellite communications.

Sky waves are severely influenced by the action of free electrons, called ions, in the upper atmosphere and are caused to be attenuated and refracted, possibly being returned to earth.

Whilst all transmitting antenna systems produce one or more of the three main modes of propagation, one of the modes will predominate. The predominant mode may be equated to the frequency used if all other constraints remain constant. For the purpose of this explanation of propagated radiowaves, it is assumed that the mode of propagation is dependent upon the frequency used, for that is the only parameter which may be changed by an operator.

Surface wave propagation

The surface wave will predominate at all radio frequencies up to approximately 3 MHz. There is no clear cut-off point and hence there will be a large transition region between approximately 2 MHz and 3 MHz where the sky wave slowly begins to predominate.

The surface wave is therefore the predominant propagation mode for the frequency bands VLF, LF and MF, which include the GMDSS NAVTEX frequency of 518 kHz and the medium frequency communications band accommodating voice communications on 2182 and DSC on 2187.5 kHz. As the term suggests, surface waves tend to travel along the surface of the earth and, as such, propagate within the earth's troposphere, that band of atmosphere which extends upward from the surface of the earth to approximately 10 km.

Diffraction and the surface wave
An important phenomenon affecting the surface wave is known as diffraction. This term is used to describe a change of direction of the surface wave, due to its velocity, when meeting a obstacle. The earth's sphere is considered to be a large obstacle to surface waves, and consequently the wave follows the curvature of the earth.

The propagated wavefront is sitting on the earth's surface or partly underground and, as a result, energy is induced into the ground. This has two primary effects on the wave: firstly, a tilting of the wavefront occurs, and secondly energy is lost from the wave. The extent of the diffraction is dependent upon the ratio of the wavelength of the frequency

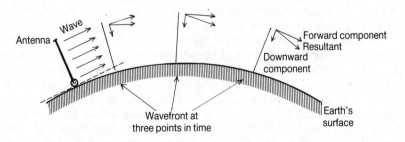

Fig. 12.6 Tilting of the surface wavefront caused by diffraction

to the radius of the earth. Consequently, diffraction is greatest when the wavelength is long (the lower frequency bands) and attenuation is greatest when the wavelength is short, causing the surface wave to predominate at the lower frequency end of the spectrum. The amount of diffraction also depends upon the electrical characteristics of the surface over which the wave travels. A major factor which affects the electrical characteristics of the earth's surface is the amount of water it contains, which in turn affects the conductivity of the ground. In practice, sea water provides the greatest attenuation of energy and desert conditions the least attenuation.

The propagation range of a surface wave for a given frequency may be increased if the power at the transmitter is increased and all other natural phenomenon remain constant. In practice however, transmitter power is strictly controlled in order to avoid interference to other users and, indeed, to limit the communication range. Figures quoting the radio range are often wild approximations. The lowest GMDSS frequency is 518 kHz used for the transmission of NAVTEX information. A NAVTEX transmitter is designed to produce an effective power output of 1 kW to provide a range of 400 miles. But, under certain conditions, NAVTEX signals may be received over distances approaching 1000 miles.

Another phenomenon caused by radiowave diffraction is the ability of a ground-propagated wave to bend around large objects in its path. This effect enables communications to be established when a receiving station is situated on the effective blind side of an island or large building. The effect is only present when LF or MF is used and, consequently, does not occur when using VHF signals. The ability of a radiowave to be diffracted around large objects is dependent upon the wavelength of the transmitted signal. In practice, the longer the wavelength of the signal in relation to the physical size of the obstruction the greater will be the diffraction effect.

Ionospheric wave propagation

The prime method of radiowave propagation in the HF band between 3 MHz and 30 MHz is by sky wave. It is by the use of frequencies in this band that global communication is achieved within the GMDSS radionet. Once again however, there is no clear dividing line between surface and sky waves. In the frequency range between 2 MHz and 3 MHz surface waves diminish and sky waves begin to predominate.

Sky waves are propagated upwards into the atmosphere where they meet ionized bands of atmosphere ranging from approximately 70 km to 700 km above the earth's surface. It is these ionized bands or layers which have a profound influence on the sky wave by returning the radiowave to earth, often over great distances.

The ionosphere
There exists above the earth's surface a number of layers of ionized energy. For the purpose of explaining the effects which the layers have on electromagnetic radiation it is only necessary to consider four of the layers. The layers are designated with respect to the earth's surface, by letters of the alphabet—D, E, F1 and F2 respectively. The ionosphere is that part of the atmosphere which extends from approximately 60 km above the earth's surface to about 800 km. Natural ultra-violet radiation from the sun strikes the outer edge of the earth's atmosphere causing an endothermic reaction to occur which, in turn, causes an ionization of atmospheric molecules. A physical change occurs which causes positive ions to be produced along with a large number of free electrons. The layers closer to the earth will be less affected than those at the outer edges of the atmosphere and, consequently, the F2 layer is more intensely ionized than the D layer. Also, the amount of ultra-violet radiation will never be constant. It will obviously

vary drastically between night and day, when the layers are on the dark side of the earth or in full daylight. In addition, ultra-violet radiation from the sun is notoriously variable during solar events and the eleven-year sun spot cycle.

Whilst it may appear that radio communication via these layers is unreliable it should be remembered that most of the environmental parameters which affect the intensity of an ionized layer are predictable. The external natural parameters which affect a layer and thus the communication range are:

- the global diurnal cycle,
- the seasonal cycle, and
- the eleven year sunspot cycle.

Radiowave ionospheric refraction
An electromagnetic radiowave possesses a wavelength, the velocity of which is affected when it passes from one medium to another of a different refractive index, causing a change of direction to occur. This change of direction is called refraction. The refractive index is a constant ratio for a pair of media and is given as:

$$n = \frac{\text{sin of incidence } \angle}{\text{sin of refraction } \angle} = \frac{\Theta_i}{\Theta_r}$$

If one of the two media is air which is not ionized, the absolute refraction index of the other medium will be obtained.

In addition to the external natural parameters which affect the ionosphere, two further controllable parameters will affect the extent of refraction of a radiowave by a given layer. These are:

- the frequency of propagation, and
- the angle of incidence of the radiowave with a layer.

Both of these parameters may be changed by a shore-based radio station, whereas the frequency can only be changed by a mobile station. When frequency is taken into consideration, the refractive index becomes:

$$n = \frac{\sin \Theta_i}{\sin \Theta_r} = \sqrt{(1 - \frac{81N}{f^2})}$$

where: N = the number of free electrons per cubic metre.
This equation demonstrates that the refractive index of a layer decreases as an inverse

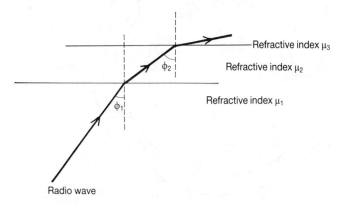

Fig. 12.7 Refraction caused by passing between layers of differing refractive index

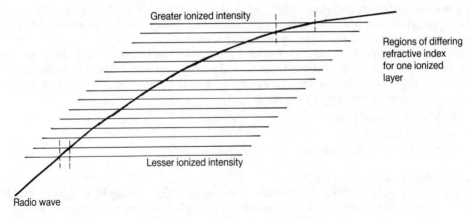

Fig. 12.8   Radiowave refraction due to progressively lower refractive indices

ratio of electron density and therefore the refractive index falls as an inverse ratio of the layer's height above the ground.

Figure 12.8 illustrates how an electromagnetic radiowave is refracted as it passes through media of progressively greater refractive index.

If, before the wave reaches the outer edge of an ionized layer the angle of incidence has reached the point where the wavefront is an right angles to the earth's surface, the radiowave will be returned to earth where it will strike the ground and be reflected back into the ionosphere.

Global communications are achieved in this way, by the radiowave making several excursions between the ionosphere and the earth's surface, each journey being known as one hop.

Because it is totally impractical for a GMDSS radio operator to calculate angles of incidence and electron densities there are a number of propagation parameters which

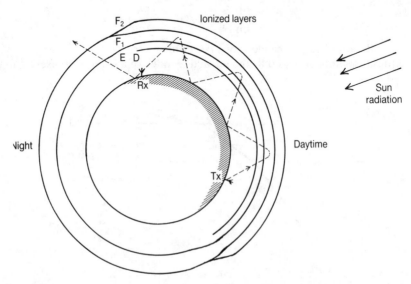

Fig. 12.9   Terrestrial global communications using multiple hops

are often quoted by global HF radio stations in order to simplify operations and the choice of frequency.

The critical frequency
There exists, for any given ionized layer, a maximum frequency that can be propagated vertically and returned to earth. As the wave is radiated vertically the angle of incidence is zero and therefore:

$$\sin 0° = 0 = \sqrt{(1 - \frac{81 N_{max}}{f^2_{crit}})}$$

$$f_{crit} = 9\sqrt{N_{max}}$$

This parameter is used purely for statistical research but it does bear a simple relationship to the important maximum usable frequency.

Maximum usable frequency
The maximum usable frequency (MUF) is the highest frequency that can be used to establish communications between two points on the earth, using a sky wave refracted from a specific ionized layer. If a higher frequency than the MUF was used the radiowave would not be refracted sufficiently to return and would pass through the layer to be lost forever. The MUF for a given layer is a function of both the critical frequency and the angle of incidence of the radiowave.

The MUF is particularly important when considering point-to-point global communications using any high frequency within the GMDSS radionet. Attenuation of a sky wave during its excursion through the ionosphere is inversely proportional to frequency. Consequently, it is essential to use the highest frequency possible for the link. However, because the MUF is dependent upon the electron density of a layer, which in turn is dependent upon ultra-violet radiation from the sun, it is not wise to use this specific frequency. The levels of ultra-violet radiation will vary causing a possible interruption in the communications link. Because of this fact, it is common to use another frequency which is approximately 15% lower than the MUF. This new frequency will provide a 15% margin of safety in operations.

Optimum traffic frequency
The optimum traffic frequency (OTF) is approximately 15% less than the MUF for a given layer. In practice, many global communications radio stations publish monthly OTF figures for specific geographical locations to enable point-to-point terrestrial communications to be reliably conducted. A second reason why the OTF chosen should be as high as possible within the frequency bands available is that of attenuation. Attenuation arises because free electrons in the electromagnetic radiowave will collide with gas molecules in a layer causing a loss of kinetic energy to occur. This, in turn, leads to a loss of wave energy.

The kinetic energy of an electron mass is:

$$\text{kinetic energy of mass} = m = \frac{1}{2}mv^2$$

Velocity is proportional to time and is therefore proportional to the periodic time of the radiowave. The periodic time is the inverse of the frequency and therefore the kinetic energy lost is proportional to

$$\frac{1}{f^2}$$

Once the OTF has been established it will be found that rarely does it fall within the designated bands available for use. The choice of frequency to be used is therefore a compromise between the OTF and those frequencies which the mobile is licensed to use.

Tropospheric wave propagation

Above 30 MHz the predominant mode of propagation of radiowaves is by the space wave. The sky wave is now rarely returned from the ionosphere because the wavelength of the carrier frequency has reduced to the point where refraction becomes insignificant.

Such a wave, when propagated upwards, passes through the ionized layers and is lost unless returned by an artificial or natural earth satellite. That wave is now called a free space wave.

If a space wave is propagated along the surface of the earth or at a short height above it, the wave will move in a straight line from transmitting antenna to receiving antenna and is often called a line-of-sight wave. In practice however, a slight bending of the wave does occur to make the radio horizon somewhat longer than the visual horizon.

The troposphere extends upwards from the earth's surface to a height of about 10 km where it meets the stratosphere. At the boundary between the two there is a region called the tropopause which possesses a different refractive index to each neighbouring layer. The effect exhibited by the tropopause on a radio space wave is to produce a downward bending action, causing it to follow the earth's curvature. The bending radius of the radiowave is not as severe as the curvature of the earth, but nevertheless the space wave will propagate beyond the visual horizon. In practice, the radio horizon exceeds the visual horizon by approximately 15%.

As most short-range communications within the GMDSS radionet are conducted using VHF it is essential that the reader is able to understand how the VHF communication range is determined.

The visual distance, d to the horizon from an antenna of height h is approximately:

$$d = \sqrt{2ah}$$

where a is the radius of the earth and h is very small in relation to it. For communications purposes the earth's radius is considered to be greater than its actual radius. A ratio a'/a is known as the effective earth radius which is quoted as 4/3.

The radio horizon now becomes:

$$d = \sqrt{2kah}$$

where $k = 4/3$

$$a' = \frac{4}{3} a$$
$$a'^2 + d_1^2 = (a + h_T)^2$$
$$d_1^2 = 2a'h_T + h_T^2$$

since a' is very much greater than h_T

$$d_1^2 \simeq 2a'h_T$$
$$d_2^2 \simeq 2a'h_R$$

the maximum range $d_{max} = d_1 + d_2$

$$d_{max} = \sqrt{2a'h_T} + \sqrt{2a'h_R}$$

a' is a constant of approximately $\frac{4}{3} \times 3960$ miles (earth radius)

$$d_{max} \text{ (miles)} = \sqrt{2h_T} + \sqrt{2h_R} \text{ (antenna height feet)}$$
$$d_{max} \text{ (km)} = \sqrt{17h_T} + \sqrt{17h_R} \text{ (antenna height metres)}$$

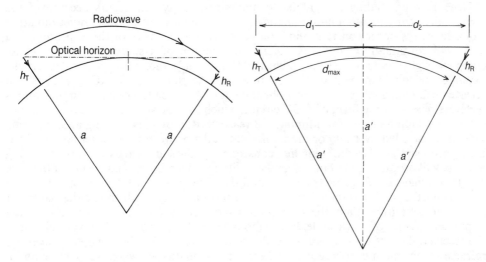

Fig. 12.10   An explanation of the 4/3 earth radius principle

The formulae above indicate that the radio range achievable for VHF communications in the GMDSS radionet is dependent upon the two main variables of transmitter and receiver antenna heights. GMDSS regulations recommend that the following formula, a variant of those above, should be used to calculate the VHF radio range in nautical miles:

$$R = 2.5 = \sqrt{h_T \text{ (metres)}} + \sqrt{h_R \text{ (metres)}} \text{ in nautical miles.}$$

Given a ships antenna height of 4 metres and a coastal radio station antenna height of 50 metres the range is 23 nmiles. This rises to 100 nmiles for antenna heights of 4 metres and 100 metres respectively. Ship to ship communications each ship having a 4 metre high antenna gives a range of 10 nmiles. Indeed SAR communications between a liferaft and another surface vessel may reach a range of only 5 nmiles. In addition VHF space waves will not pass through, or be diffracted around, large objects such as buildings or islands, in their path. This gives rise to extensive radio blind areas behind large objects

Signal fading

One of the major difficulties encountered when radiowaves are propagated via the earth's atmosphere is that of signal fading. Fading is a continual variation of signal amplitude experienced at the antenna input to a receiving system. In practice, fading may be random or periodic but in each case the result will be the same. If the signal input to a receiver falls below the quoted sensitivity figure there may be no output from the demodulator and hence the communications link is broken. If the signal amplitude at the antenna doubles, a large increase in audible output will be produced either causing discomfort to the operator or overloading automatic systems. Whatever the cause, steps must be taken at the receiver to overcome the problem of signal fading. Fading may be classified as one of three main types:

- general signal fading,
- selective fading, or
- frequency selective fading.

General signal fading in a global communications system occurs because of the continually changing attenuation factor of an ionospheric layer. Ultra-violet radiation from the sun is never constant and, consequently, the intensity of the ionization of a layer will continually change. The signal attenuation of a specific layer may cause complete signal fade-out as the intensity of the sun's radiation changes. With the exception of this extreme case, the use of automatic gain control (AGC) circuits in a receiver will effectively combat this phenomenon. All broadcast and communications receivers possess some form of AGC circuit, which may be very sophisticated.

Selective fading occurs for a number of reasons. Radiowaves arriving at an antenna may have travelled over two or more different paths between transmitter and receiver. Each path-length will be different and, consequently, the signals arriving at the receiving antenna will produce a combined signal amplitude, which is the phasor sum of the two. The two signals, of the same frequency and the same origin, will be out of time phase with each other and will therefore produce a resultant signal which is either larger or smaller in amplitude than the original. In most cases the signal path lengths are unpredictable and often variable, leading to the need for a good quality AGC circuit in the receiver. This effect can occur, as shown in figure 12.11, when two sky waves are refracted from the ionosphere over different path lengths, or when a sky wave and a ground wave are received together, or when two ground waves are received over different paths.

Frequency selective fading occurs where one component of a transmitted radiowave is attenuated to a greater extent than other components. In any radio-telephone communications link a large number of frequencies are contained within the bandwidth

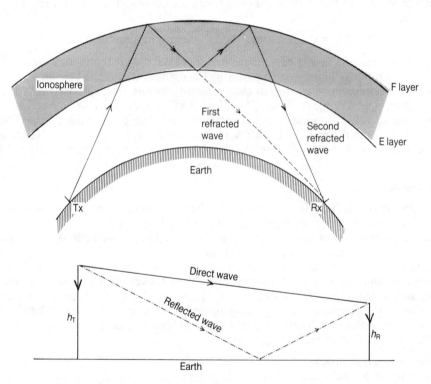

Fig. 12.11   Signal fading caused by multipath effects

of the transmitted signal. The individual frequencies contained in the transmission are those of the fundamental carrier frequency plus the RF frequencies generated by the method of modulation employed. To produce an error free or distortion free radio-telephone link all modulation baseband frequencies at the transmitter must be faithfully reproduced at the receiver output. If any of the modulated frequencies are lost in the transmission medium, which may happen when frequency selective fading is present, they cannot be reproduced at the receiver.

More importantly however, if the carrier frequency is lost in the transmission medium it will be impossible to demodulate the audio intelligence at the receiver, unless specific circuitry is available and the carrier loss is predictable.

Frequency selective fading cannot be cured by the use of AGC circuits in a receiver. Its effects can, however, be limited by using:

- a transmission which radiates one mode only (Morse A1A),
- single sideband (SSB,J3E) suppressed carrier transmission, or
- frequency modulation.

Radio-telephony communications within the GMDSS radionet utilize SSB amplitude modulated signals on the MF and HF bands and frequency modulation of the VHF band.

13
Antenna systems

13.1 Introduction

An antenna is arguably the single most critical part of any radio communications system and those to be found in the GMDSS radionet are no exception. Unfortunately, it is often the part of a radio installation which is less than efficient. Not because of the characteristics of antenna design but because of the major problems of antenna siting and installation, particularly on modern ships operating in the GMDSS system.

As ships become more streamlined, the available antenna space reduces, often to the point where multiple antenna systems simply cannot be fitted.

As an example, where for instance would an Inmarsat-A or B steerable satellite antenna be mounted on an aircraft or a truck where space is very critical? Similarly, consider the need to stabilize gyroscopically and move satellite antenna systems mounted on aircraft or ships. The MF band used for terrestrial communications also possesses problems. Long wire antennae, needing to be approximately 150 m (half wavelength at 2 MHz) are virtually impossible to erect on modern ships.

GMDSS communications equipment uses a variety of antennae, each one designed with individual characteristics to suit operational needs. Dipole, whip and parabolic antennae are the most common, but whatever the construction they all operate using similar principles.

Figure 13.1 shows the efficient antenna farm of a modern vessel including GMDSS, navigational and entertainment antennae.

The following description of antenna systems is limited to that needed to understand radio communications systems and operation within the GMDSS radionet. Whilst basic antenna theory is considered, it should be noted that it is only necessary for the reader to understand antennae from an operational and installation point of view.

An antenna is essentially a piece of wire which may or may not be open at one end. The shortest length of wire that will resonate at a single frequency is one which is critically long enough to permit an electric charge to travel along its length and return in the period of one cycle of the applied radio frequency. The period of one cycle is called the wavelength. The velocity of an RF charge is that of light waves, $299,793,077 \text{ m s}^{-1}$, which is approximated to $300 \times 10^6 \text{ m s}^{-1}$ for convenience. The wavelength of any RF wave is therefore:

$$\lambda = \frac{300 \times 10^6}{f}$$

Because the RF charge will travel the length of the wire and return, it follows that the shortest resonant wire is one half of a wavelength long. In fact many GMDSS antenna systems are called half-wave or $\lambda/2$. If, as an analogy, the resonant length is assumed to be a trough with obstructions at each end and a ball is pushed from one end, it will strike the far end and return, having lost energy. If, at the instant the ball hits the near end

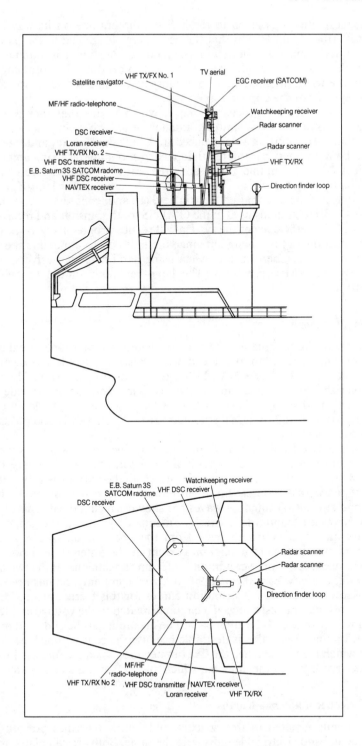

Fig. 13.1 GMDSS antenna installation aboard the 'Eliza PG'. (Courtesy *Ocean Voice*)

obstruction, more energy is given to the ball it will continue on its way indefinitely. However, it is critically important that the new energy is applied to the ball at just the right time in order to maintain the action. In practice, if the timing is found to be in error the length of the resonant trough may be changed to produce the optimum transfer of energy along the wire. antennae therefore, must be constructed to a critical length to satisfy the frequency of the applied RF energy.

Antennae, exhibit the reciprocity principle, which means that they are equally as efficient when working as a transmitting antenna or as a receiving antenna. The main difference is that a transmitting antenna needs to handle high power and is usually more substantially built and better insulated than a corresponding receiving antenna.

For efficient radio communications, both the transmitting and receiving antennae should possess the same angle of polarization with respect to the earth. Polarization refers to the angle of the transmitted electric field (E) and, consequently, if the E field is vertical, as in the VHF systems used in the GMDSS, both transmit and receive antennae must be vertical. The efficiency of the system will reduce progressively as the error angle between transmitting and receiving antennae increases up to a maximum error of 90°. The reader should remember this fact when using hand-held transceivers, particularly during distress and SAR operations. The hand-held unit should be held with the antenna vertical.

Half wavelength antenna

An antenna operating at precisely half a wavelength is traditionally called a Hertz antenna. Many antennae do not operate at $\lambda/2$ because they would be excessively long. Take, for instance, the GMDSS NAVTEX system which operates on 518 kHz with a vertically polarized transmitting antenna. A receiving antenna to provide maximum efficiency would need to be 289.375 m high. Clearly an unacceptable antenna. As a consequence, most MF and some HF antennae are both electrically and physically short and therefore less efficient.

A $\lambda/2$ antenna is effectively a $\lambda/4$ transmission line with a signal generator, the transmitter, at one end and an open circuit at the other, as shown in figure 13.2.

Ohm's Law states that when an open circuit exists the current will be zero and the potential difference (p.d.) across the open circuit will be maximum. In figure 13.5 voltage (E) and current (I) standing waves are shown which indicate this fact. E and I distribution curves are standard features of antenna diagrams. If the generator (signal source) is one-quarter wavelength back from the open circuit, the E and I curves indicate minimum voltage and maximum current at the antenna feed point. In most cases this is the desirable E and I condition for feeding an antenna. If the two arms of the transmission line are now bent through 90°, a $\lambda/2$ efficient antenna has been produced. Figure 13.5 shows the E and I distribution curves for this Hertz antenna. Once again Ohm's Law states that the resistance of a circuit is related to the voltage and the current. In this case the impedance of the antenna will be maximum at the ends and minimum at the centre feed point. Again this is desirable because the centre impedance is approximately 73 Ω which ideally matches the 75 Ω impedance coaxial cable used to carry the output of the transmitter to the antenna, which may be some distance away.

Physical and electrical antenna lengths

Ideally, an antenna isolated in free space would follow the rules previously quoted whereby the actual and electrical lengths were the same. Both are calculated to be $\lambda/2$ of the transmission frequency. However, because the velocity of the radiowave along the

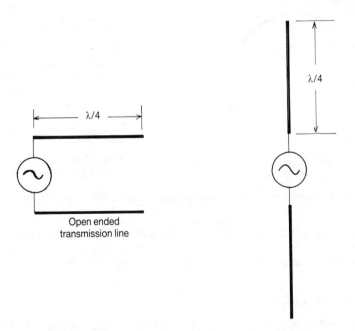

Fig. 13.2   Half-wavelength antenna derived from a quarter-wavelength transmission line

wire antenna is affected by the antenna supporting system and is slightly less than that in free space, it is normal to reduce the physical length of the antenna by approximately 5%. The corrected physical length of an antenna is therefore 95% of the electrical length.

Antennae and feeders are effectively matched transmission lines which, when a radio frequency is applied, exhibit standing waves, the length of which is determined by a number of factors outside the scope of this book. However, the waves are basically produced by a combination of forward and reflected power in the system. A measurement of the ratio between forward and reflected power, called the standing wave ratio (SWR), provides a good indication of the quality of the feeder and the antenna. Measurement of the SWR is made using voltage and becomes VSWR.

Antenna radiation patterns

A graph showing the actual intensity of a propagated radiowave at a fixed distance as a function of the antenna system is called a radiation pattern. Most antenna radiation patterns are compared with that of a theoretical reference antenna called an isotropic radiator. Radiation patterns may be shown as the H plane or the E plane of transmission. Figure 13.3 shows the E-plane radiation patterns of an isotropic radiator and a λ/2 dipole antenna. It should be noted that this is a two-dimensional diagram whereas the actual radiation pattern will be three-dimensional, as shown in figure 13.6, and affectionately known as the 'doughnut' pattern.

The maximum field strength for the λ/2 dipole occurs at right angles to the antenna and there is very little radiation at its ends. This type of antenna in the horizontal plane is therefore directional, whereas the isotropic radiator is omnidirectional. However, the λ/2 antenna is omnidirectional when used vertically polarized in the GMDSS radionet.

A second important principle of an antenna is its beamwidth. The radiation pattern is

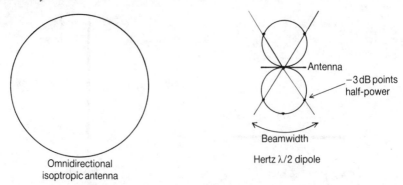

Fig. 13.3   Two-dimensional radiation patterns for an omnidirectional antenna and a Hertz antenna

able to illustrate the antenna beamwidth which is calculated at the 'half power points' or −3 dB down from the peak point. If the receiving antenna is located within the beamwidth of the transmitting antenna good communications will be made.

Antenna gain and directivity

Antenna gain and directivity are very closely linked. The greater the directivity an antenna exhibits, the greater it will appear to increase the transmitted signal in a specific direction. The $\lambda/2$ dipole, for instance, possesses a gain of typically 2.2 dB, on those planes at right angles to the antenna, when compared with an isotropic radiator. As a consequence, there will be zero signal propagated along the other two planes in line with the dipole. Both properties of gain and directivity are reciprocal and apply equally to both transmitting and receiving antennae.

As an example of gain, if a GMDSS VHF transmitter with a peak output power of 25 W was correctly matched to a directional antenna with a gain of 3 dB, the effective isotropic radiated power (EIRP) would be double, thus increasing the field strength at the receiver. It should be remembered however, that the antenna would be directional.

Practically, it is important to consider the effect of both the transmitter and receiver antenna gains in a complete radio communications system. The formula below provides a simple method of calculating the signal strength at a receiver input:

$$P_r = \frac{P_t G_t G_r \lambda^2}{16\pi^2 d^2}$$

where; P_r = power received in watts,
   P_t = power output of transmitter in watts,
   G_t = the ratio gain of the transmitting antenna,
   G_r = the ratio gain of the receiving antenna,
   λ = wavelength of the signal in metres,
   d = the distance between antennae in metres.

Ground effects

The overall performance of an antenna system is extensively changed by the presence of the earth beneath it. The earth acts as a reflector and, as with light waves, the reflected radiowave leaves the earth at the same angle with which it struck the surface. Figure 13.4 shows the direct and reflected radiowaves from a transmitting antenna.

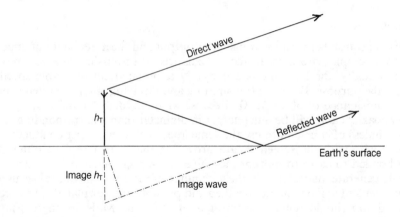

Fig. 13.4 Direct and earth reflected radiowaves from a transmitting antenna form a sky wave

Because the surface of the earth is rarely flat and featureless, there will be some directions in which the two waves are in phase, and thus are additive, and some directions where the two are out of phase, and thus subtractive. This effect gives rise to a change in the directive properties of the antenna system and, in fact, means that it is virtually impossible to create a perfect omni-directional antenna.

An antenna design known as the ground-plane antenna uses an artificial ground assembly instead of the earth to achieve its propagational characteristics.

The antenna is formed of a driven vertical quarter-wavelength antenna with an artificial metallic ground plane constructed of four rods perpendicular to the antenna (horizontal to the earth) extending radially from the base. This array is used for communication in the HF bands below 30 MHz because, unlike the unmodified dipole, it will provide radiation at a low angle.

Additionally, the ground effect can be used to advantage with other systems. If a hand-held GMDSS VHF transceiver is placed with its base in contact with a metal plate above the earth it is possible to increase the effective radio range but at the expense of directivity.

Antenna efficiency

Efficiency is of particular importance to all communications systems. If the efficiency of a GMDSS VHF antenna drops to 50%, the 25 W permitted maximum radiated output could drop to about 12.5 W, leading to a consequent loss of range. It would be rare to find any system which is 100% efficient and antennae are no exception. However, antenna losses are well documented and, consequently, the EIRP figure for a system is usually calculated with reference to known efficiency figures. Losses leading to inefficiency in an antenna system may generally be classed as dielectric losses affecting the transmission properties of the antenna. Such losses in a transmitting antenna, may be produced by arcing effects or corona discharge whereby bad connections or wiring create a loss of energy into the atmosphere. This phenomenon is easily seen, particularly at night. Other losses are produced by inducing small eddy currents into neighbouring metal objects such as the superstructure of a ship or its funnel.

Most of these losses can be eliminated by careful wiring and good positioning of the antenna.

Antenna feed lines

Whilst the connection between the transmitter output and the antenna input appears to be made by a simple wire it is, in fact, a balanced transmission line and possesses impedance. Usually, the feed line is a correctly terminated coaxial cable specifically designed for the purpose. For most transmitting and receiving antenna systems the feed line has an impedance of 50 or 75 Ω. Because of its need to handle more power, a transmitter coaxial cable will be much larger in diameter than a corresponding receiver coaxial line, unless of course both use the same line. The inner copper conductor forms the live feed wire with the screen sheath providing the earth line. The outer sheath should be bonded to earth to prevent inductive pick-up in the centre conductor wire, which would generate interference in the communications link. Coaxial cables used in a marine environment will be double sheathed and possibly armour plated. They are fully waterproofed and should remain so throughout their life. Moisture ingress into the cable insulation material will cause considerable losses as energy is absorbed and not radiated.

The Marconi antenna

The Marconi antenna may occasionally be found in GMDSS installations, particularly on receivers operating in the MF band or below. The main difference between this antenna and those previously discussed is that the Marconi antenna is invariably $\lambda/4$ long and requires a good conducting path to earth in order to operate efficiently.

Figure 13.5 illustrates that this antenna possesses similar properties to the $\lambda/2$ dipole with half of its length being reflected beneath the earth. Virtually the only difference between the two antennae is that the Marconi antenna has a feed point impedance of approximately 36 Ω and consequently needs some form of impedance matching arrangement when coupling it to a feeder.

This antenna relies for its efficient operation on the conductive properties of the ground. If those properties are less than perfect an earth conductor or screen may be used, particularly where the antenna is used above the ground on top of a building or other structure.

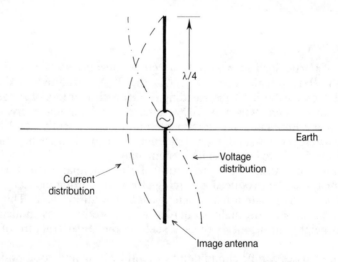

Fig. 13.5   A grounded Marconi antenna showing the image antenna

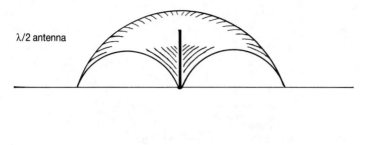

Fig. 13.6 VHF vertical antenna horizontal radiation patterns

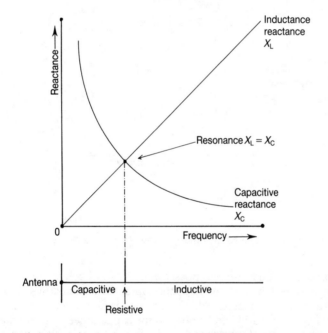

Fig. 13.7 Graph of change of inductive reactance and capacitive reactance with frequency

Loaded antennae

An antenna must be resonant at the frequency of transmission or reception in order to be effective. In many cases $\lambda/2$, $\lambda/4$ or other sub-divisions are used to make the antenna length match the wavelength of the transmission and thus be resistive. If a tuned circuit is not at resonance with the frequency applied its reactance may be inductive or capacitive.

At the low frequency (long wavelength) end of the spectrum, antennae tend to be too short and exhibit capacitive effects. It is essential to reduce this capacitive effect by changing the electrical characteristics of the physical antenna assembly.

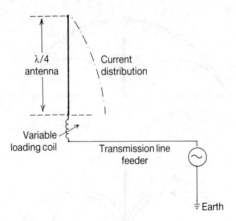

Fig. 13.8 Diagram of a base loaded Marconi antenna

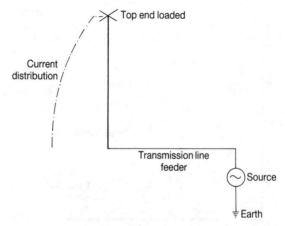

Fig. 13.9 Current distribution of a capacitive top-loaded Marconi antenna

Figures 13.8 and 13.9 illustrate two possible methods of achieving this.

Additional capacitance may be added to the top of the antenna in such a way that the overall capacitive effect can be reduced. The antenna is now termed top-loaded. The top end assembly in figure 13.9 adds additional shunt capacitance to ground and thus is in series with the antenna actual capacitance, reducing the overall effective capacitance. A second way of achieving the same effect is to include a coil at the base of the antenna in series with the feeder line. This has the effect of increasing the inductive part of the system, thereby decreasing the capacitive effect. This antenna is known as a bottom-loaded antenna.

If, in a receiving system, the modified antenna assembly is still not effective it is likely that the antenna will be made active.

Active antennae

An active antenna, such as the type used for **GMDSS NAVTEX** reception, is one which has an in-built signal amplifier at the base of the physical antenna. This amplifier provides a high degree, typically 10 dB, of signal gain before the signal passes down the coaxial feeder wire where losses will occur. It is essential that the amplifier is fitted directly at the base of the antenna in order to amplify the signal before external noise degrades the signal/noise ratio further. The coaxial feeder will therefore carry a d.c. power supply up the mast to supply the amplifier whilst coupling the signal down to the receiver input.

13.2 Specialist antennae

There are many types of antenna assemblies, or antenna arrays, used to form specific functions in radio communications. A number of them require complex assemblies constructed over huge areas of ground and are thus used in GMDSS shore-based installations only. Two of these are the rhombic and the log-periodic arrays. Another huge antenna construction is the satellite parabolic antenna which forms the ground-satellite up/down link for satellite services. A small version of this will be found on an Inmarsat fitted ship or other MES. In addition, the highly directional Yagi-Uda array may be found in GMDSS VHF installations. Other antennae include the long-wire, the dipole or whip antenna and the omnidirectional collinear. Full technological details of all communications antennae may be found in one of the numerous books dedicated to the subject. However, a brief description of the mobile antennae used in the GMDSS is included here.

The whip antenna

The term is used to describe a flexible vertical wideband antenna which is used to communicate on a range of frequencies. Typical wide-band frequency ranges are 100 kHz to 30 MHz for those fittings which include Morse telegraphy or more commonly 1.4 MHz to 30 MHz in the GMDSS. The mechanical length of the antenna may be in the range 5 m to 12 m depending upon the requirements of the installation. Electrical lengths range from 4.4 m to 10.6 m.

Obviously, if a fixed-length antenna is to be used over such a wide band of frequencies it cannot be critically resonant at all frequencies. This fact is true of vertical whip antennae used in the GMDSS. They are, therefore, less than perfect radiators or receivers but possess the advantage that the antenna does not have to be changed when changing frequencies. A whip antenna may be considered to possess omnidirectional properties.

The vertical half-wavelength dipole antenna

A vertically mounted half-wavelength dipole is used for VHF communications. Again it is a wideband antenna although the frequency band is somewhat smaller than at MF or HF. The maritime VHF band is 156 MHz to 162 MHz although in the GMDSS the VHF band may extend to 144 MHz to 165 MHz.

Again it is obvious that the antenna cannot be resonant on all the frequencies within the band. In practice, it is made to be resonant at the centre of the operational frequency band with a consequent reduction in efficiency the further the frequency deviates from the centre.

The collinear antenna

Another VHF antenna which may be encountered is the multi-element driven collinear. Depending upon its construction, this type of antenna may provide up to 6 dB

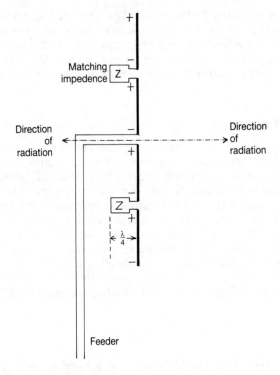

Fig. 13.10   A four-element collinear array

of gain in the system. A collinear array is constructed of any number of half-wavelength elements end to end in a linear pattern. Each element of the array is driven in order to produce individual radiation fields which are all in phase and thus additive at points perpendicular to the array.

Gain is increased by adding more half-wavelength elements but at the expense of directivity. The omnidirectional collinear antenna possesses a radiation pattern similar to the doughnut pattern of the Marconi antenna but much flatter.

The Yagi-Uda antenna

This antenna will be readily recognized as that used to receive television pictures ashore but may also be used for VHF communications work. The antenna consists of a half-wavelength driven element plus the addition of added parasitic elements. The complete array is highly directional.

The vertical driven element alone would possess an omnidirectional radiation pattern. This pattern is modified by the addition of parasitic elements placed at 0.1λ wavelength from the driven element. The reflector, placed behind the driven element, is, at 0.55λ slightly longer than the active element, and has the effect of reducing reverse propagation to a minimum. The director, placed in front of the driven element, is, at 0.45λ slightly shorter than the active element, and directs the radiation pattern along the antenna axis. Gain is thus achieved by adding more directors to shape the radiation pattern along the axis. However, as gain is increased the beamwidth is decreased, thus creating greater directivity.

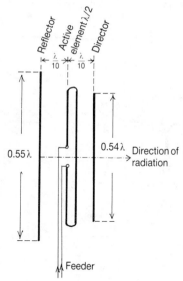

Fig. 13.11   A three element Yagi-Uda array

Microwave parabolic antenna

Ship earth stations' above deck equipment (ADE) includes a parabolic reflector antenna popularly known as the dish antenna. This antenna format is used because it is able to contribute a high level of signal gain to both the transmit and receive signal levels. Whilst there are a number of types of parabolic antenna, including the offset dish used for domestic satellite television reception, they all operate using the same principles. A SES uses a prime focus system, whereby the received signal from the satellite is reflected, by the perfect parabola formed by the dish, to be collected at the prime focus point. The antenna works in the reciprocal way to this when used on transmit.

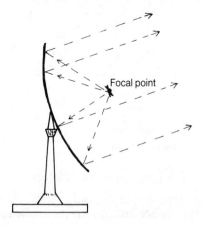

All radiated energy is concentrated in a beam by the reflector of this parabolic antenna. The reciprocal action occurs when signals are received from a satellite

Fig. 13.12   A prime focus parabolic antenna

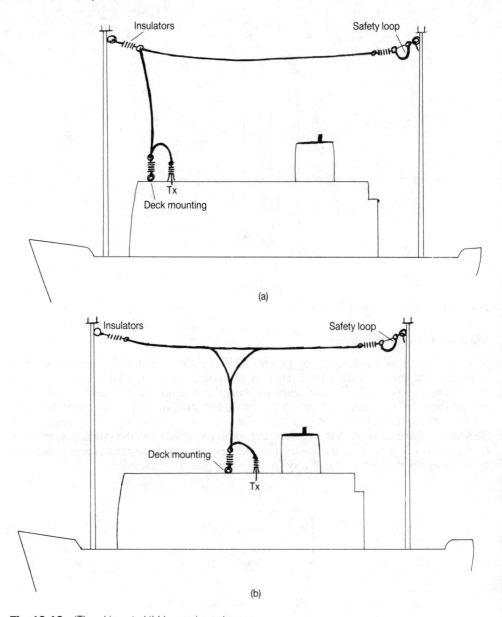

Fig. 13.13   'T' and inverted 'L' long-wire antennae

The gain of the antenna is dependent upon two factors:

(a) the wavelength of the received frequency, and
(b) the diameter, and thus the surface area, of the parabolic reflector.

The signal wavelength cannot be changed by the design engineer, but the size of the reflector can.

Typical antenna gain for an SES parabolic antenna of 0.9 m diameter operating at 1.6 GHz is:

$$\text{Signal wavelength} = \lambda = \frac{3 \times 10^8}{1.6 \times 10^9} = 0.1875 \text{ m}$$

$$\text{Gain} = 6 \left(\frac{D}{\lambda}\right)^2 = 6 \left(\frac{0.9}{0.1875}\right)^2 = 138.24$$

Antenna beamwidth is also a function of reflector diameter.

$$\text{Beamwidth} = 70 \frac{\lambda}{D} = 70 \left(\frac{0.1875}{0.9}\right) = 14.58°$$

Consider reducing the diameter of the ADE dish by half to 0.45 m in order to reduce the size of the ADE assembly. The gain offered by the parabolic antenna would now fall to 14.4, or a virtual fall of 90%. An Inmarsat-A SES given this level of antenna gain is unlikely to meet the Inmarsat signal specifications for transmit and receive.

The long-wire antenna

This type of antenna used to be a standard fitting on merchant ships because it provided excellent radiation properties at the medium frequencies used for communications. It was relatively easy to install this type of antenna on ships which were designed with two high masts—a feature which has been lost as ship design has been streamlined. However, long-wire antennae are still fitted where possible in order to satisfy the wavelength requirements of medium frequency communication. At 2 MHz a half-wavelength antenna would need to be 150 m long in order to be efficient. Clearly this is totally impracticable when considering a vertical whip antenna. But with long-wire antennae, lengths approaching 75 m (quarter wavelength) are possible, which with some electrical adjustments produce a good radiator. Two types of long-wire antennae may be found; the ' T ' assembly and the inverted ' L ', the names of which become obvious by reference to figure 13.13.

An antenna is considered to be a long wire only when it is long in terms of wavelength. In practice, the long-wire antenna on board a ship will probably be short compared with the wavelength of the transmission frequency. The transmitter feeding the antenna will therefore have an arrangement of variable inductance and/or capacitance selected automatically in order to satisfy the electrical requirements of the antenna. When erected high above the waterline this antenna offers considerable power gain over the dipole antenna. In practice, the longer the antenna the greater the gain. The antenna is directive, with maximum radiation occurring off the ends of the assembly and little broadside to it. The physical length of a long-wire antenna needs to be slightly (0.05%) less than the electrical length.

14
Radiocommunications systems

14.1 Introduction

Terrestrial communication within the GMDSS involves the transfer of information which exists in one or more different modes—voice, telex (telegraphy), facsimile and data. The raw baseband signals, for these modes, are modulated onto the carrier frequency for propagation, using one of a number of modulation methods—amplitude modulation (AM), frequency modulation (FM) or phase modulation (PM). As with all electronic systems, valued judgements must be made when considering the relative advantages and disadvantages of using one modulation system in preference to another. The design choice of the modulation method to be used in a communications system is based on a number of factors, but for the purpose of this book it is essentially the signal power, the bandwidth and the noise which are to be considered.

Voice and facsimile communications, when compared with telex and data, tend to be very slow in the rate at which information can be transferred. Voice communications is a good example. This slow, unreliable, bandwidth consuming method of communications is desirable when human beings are conversing with each other. The disadvantages of voice communications are therefore hugely outweighed by the convenience factor. In addition, voice communications can be made by totally unskilled operators whereas telex and data communications require the operator to be keyboard literate.

14.2 Voice communications

Voice communication between two stations must be as close to normal speech communication as possible. Quality is of paramount importance. The receiver must faithfully reproduce the baseband speech signal presented to the microphone in the transmitter a great distance away. Language, dialect and intonation must be communicated if the communications link is to be perfect. However, even a perfect voice communications link holds disadvantages. Two people from different countries may use different languages, hence the use of a common standard language—English—in the GMDSS. People from the same country may misunderstand each other because of differences in dialect or speech intonation. In practice, the human voice baseband signal undergoes considerable processing before being applied to the modulator of a transmitter. Depending upon the modulation system used for the carrier frequency selected, the baseband voice signal may be frequency band limited, amplitude limited, pre-emphasized, companded or ultimately encoded or digitized. All this is done in an effort to maintain voice quality whist minimizing the bandwidth used on transmission. Whatever form the baseband signal processing takes before the modulator in a transmitter the precise reciprocal action must take place in the audio processing circuitry of the receiver in order to reconstitute the audio signal faithfully.

It is, however, understood that voice communication is the most easily used and convenient method of information transfer and, as such, will always be used in the GMDSS.

Signal modulation and bandwidth

Amplitude modulation

Amplitude modulation of the baseband signals onto a defined carrier frequency is a method which has been used for many decades because the process is relatively simple and because baseband signals can be readily reconstituted in the receiver using very simple demodulator circuitry. This simplicity makes AM very popular for entertainment broadcast radio systems where very cheap receivers can be produced. However, the simplicity is achieved at the cost of increased noise and, more importantly, the requirement for increased bandwidth usage in the frequency spectrum. However, AM will continue to be used in terrestrial services and indeed is used in the GMDSS for MF/HF global communications.

Amplitude modulation of a carrier wave signal is achieved by varying its instantaneous amplitude at the rate of change of amplitude of the audio wave.

The mathematical expression for a sinusoidal carrier wave is:

$$v = V_c \sin (\omega_c t + \theta)$$

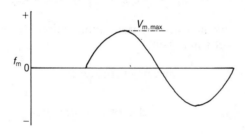

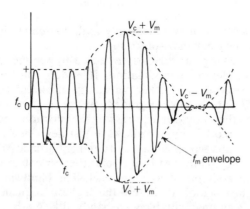

The amplitude of the carrier frequency (f_c) is increased to 2 V_c and reduced to zero by the modulating signal

Fig. 14.1 A 100% amplitude modulated waveform

where: v = the instantaneous carrier voltage,
  V_c = the peak amplitude of the carrier voltage
  ω_c = 2 π times the carrier frequency, and
  θ = is the phase of the carrier voltage at time $t = 0$.

The phase of the carrier is not shifted and consequently θ is considered to be zero. The expression for the instantaneous voltage amplitude of the modulating signal is:

$$v + V_m\sin\omega_m t$$

where: V_m = the peak value
  $\omega = 2\pi \times$ the freqency

Carrier amplitude should vary sinusoidally about the peak value of the carrier voltage V_c. The frequency of the variation will be that of the audio signal $\omega_m/2\pi$ in Hertz whereas the peak variation will be V_m.

The amplitude of the modulated carrier wave is now $V_c = V_m\sin\omega_m t$ giving an instantaneous voltage of the modulated radiowave:

$$v = (V_c + V_m\sin\omega_m t)\sin\omega_c t$$
$$v = V_c\sin\omega_t + V_m\sin\omega_m t\sin\omega_c t$$

By using trigonometry the defined equation becomes

$$v = V_c\sin\omega_c t + \frac{V_m}{2}\cos(\omega_c - \omega_m)t - \frac{V_m}{2}\cos(\omega_c + \omega_m)t$$

and illustrates that the basic method of amplitude modulation creates a transmitted complex wave with three distinct components:

$$(a)\, f_c = \omega\,\frac{C}{t}\ \text{the carrier frequency,}$$

$$(b)\, f_c - f_m = \frac{\omega_c - \omega_m}{2\pi}\ \text{the lower sideband,}$$

$$(c)\, f_c + f_m = \frac{\omega_c + \omega_m}{2\pi}\ \text{the upper sideband.}$$

Complex waves are difficult to show graphically and, consequently, it is usual to illustrate AM signals using the frequency spectrum.

The maximum amplitude change of the carrier caused by the modulating signal is 100% and is known as the depth of modulation. It is desirable to achieve 100% modulation of the carrier because any smaller percentage of modulation would cause the signal-to-noise ratio (S/N) to become degraded. But depths of modulation above 100% should not be used as they would introduce severe distortion (see figure 14.1).

The method of amplitude modulating a radiowave just described creates a double sideband (DSB) radiowave of the type used on entertainment broadcast radio networks complete with carrier frequency. DSB AM is not permitted for communications in the GMDSS terrestrial MF/HF communications radionet. The bandwidth of the transmitted signal created by this method is double the highest frequency of the audio modulating waveform giving an unacceptable wide bandwidth. As the upper and lower sidebands are identical in content it is wasteful of both power and spectrum bandwidth to transmit both. For maritime MF/HF communications work the lower sideband is suppressed to produce upper single sideband (SSB) modes, thus halving the channel bandwidth requirements with a consequent improvement in spectrum efficiency.

There are three operational modes of SSB in use. They are labelled by

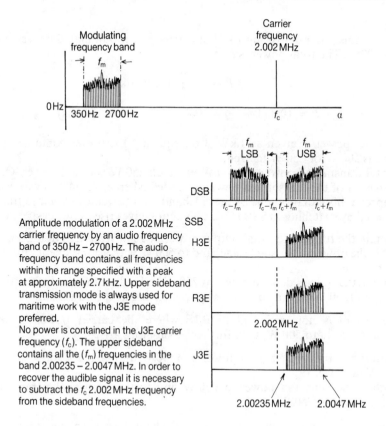

Amplitude modulation of a 2.002 MHz carrier frequency by an audio frequency band of 350 Hz – 2700 Hz. The audio frequency band contains all frequencies within the range specified with a peak at approximately 2.7 kHz. Upper sideband transmission mode is always used for maritime work with the J3E mode preferred.

No power is contained in the J3E carrier frequency (f_c). The upper sideband contains all the (f_m) frequencies in the band 2.00235 – 2.0047 MHz. In order to recover the audible signal it is necessary to subtract the f_c 2.002 MHz frequency from the sideband frequencies.

Fig. 14.2   DSB/SSB AM

international convention depending upon the transmitted power contained in the sideband and/or the carrier:

- H3E—a single upper sideband containing the signal intelligence plus a carrier frequency of maximum power amplitude.
- R3E—a single upper sideband as above plus a carrier frequency of reduced power amplitude.
- J3E—a single upper sideband as above plus a fully suppressed carrier.

As AM creates a complex waveform essentially containing three elements, the carrier wave and the two sidebands, the total peak envelope power (PEP) radiated by a transmitter must be divided between the three. It should be remembered that the power contained in the carrier wave is lost during the demodulation process at the receiver and, as a consequence, is a waste to the system. The formula below is used to illustrate the power changes that occur within the three elements as the modulation mode is changed.

The power developed by an AM radiowave is the sum of the powers of all the frequencies present in the wave. The total power is:

$$p_t = P_c(1 + \frac{m^2}{2}) \quad \text{(watts)}$$

Example

Consider a transmitter delivering 1 kW of carrier power modulated by a sine wave to a depth of 100%. The total power is:

$$P_t = P_c(1 + \frac{m^2}{2}) \quad \text{(watts)}$$

$$P_t = 1000(1 + \frac{1}{2}) = 1000 \times 1.5 = 1500 \quad \text{(watts)}$$

The carrier power is given as 1 kW. Consequently the power contained in the two sidebands is 500 W or 250 W in each.

The total transmitted power is 1500 W of which 250 W is useful power. Or, in other words, 16.66% of the transmitted power is needed whereas 83.34% is wasted.

To improve this situation one of the sidebands can be removed before transmission, as is achieved by switching to SSB H3E. This action gives rise to two design possibilities:

(a) maintain the transmitter total output power at 1500 W and, in doing so, effectively double the power in the single sideband to 500 W with a consequent improvement in S/N ratio or,
(b) maintain the single sideband power at 250 W and reduce the transmitter total output power to 1250 W with a consequent power saving at the power supply.

If the system is now switched to R3E where the carrier power is reduced by approximately 50% to 500 W, then further possibilities arise:

(a) maintain the transmitter total power at 1500 W and create a single sideband with 1000 W of power or,
(b) maintain the sideband power at 250 W and further reduce the transmitter total output power to 750 W.

Finally, if the system is now switched to J3E where the carrier power is considered to be non-existent the following cases arise:

(a) maintain the transmitter total power at 1500 W with all this power contained in the single sideband or,
(b) maintain the sideband power at 250 W and reduce the transmitter total power to this figure.

It is also possible, using this form of signal modulation, to produce two independent sidebands (ISB) whereby two totally separate intelligence signals are modulated onto the same pilot carrier frequency. The system is widely used in military applications and is slowly becoming popular in commercial systems.

The advantages of using SSBSC (H3E) in preference to DSB

(a) The bandwidth on transmission is half of that needed for DSB and, consequently, twice the number of channels may be created on the limited frequency spectrum available for communications.
(b) The S/N ratio is greatly improved. Noise power on the frequency spectrum is proportional to the bandwidth utilized and, consequently, half the bandwidth relates to half the noise interference. Also, as has been shown, there will be a greater amplitude of power contained in the sideband to overcome the effects of noise by improving the S/N ratio.
(c) The power efficiency of the transmitter improves. When using DSB the transmitter power amplifier is operating continuously even when no intelligible communication

is being made. When using J3E, power will only be radiated from the transmitter when the sideband is present.
(d) Selective fading of the signal is unlikely to be a problem. The wider the bandwidth of the mode in use the greater will be the problem with selective fading.

Frequency modulation
Frequency modulation occurs when the instantaneous frequency of a carrier wave differs from the carrier frequency by an amount proportional to the instantaneous amplitude of the modulation wave, whilst the amplitude of the carrier wave remains constant.

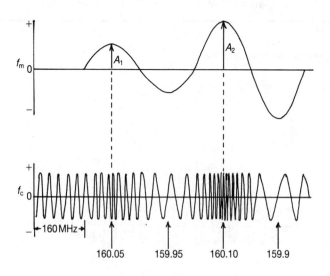

The frequency shift of the carrier wave doubles as the amplitude of the modulating signal doubles

Fig. 14.3 Waveform showing frequency shift caused by amplitude changes

It is clear from figure 14.3. that when the modulating signal reaches its maximum positive value the modulated carrier wave reaches its maximum increased frequency shift, and vice versa in the negative direction. Because the frequency shift of the carrier wave is dependent upon an amplitude shift of the modulating wave, some confusion may exist between AM and FM. Figure 14.4 summarizes the major differences between AM and FM.

It should be noted that the two characteristics of the audio modulating wave— amplitude change and frequency change—are both represented in the frequency modulated carrier wave. If a 1 kHz (1 ms per cycle) pure tone signal is used to modulate a 1 MHz carrier wave, the result will be a change of frequency caused by an amplitude change of the modulating signal occurring in a time period caused by the frequency change of the modulating signal.

If the amplitude of the modulating signal is doubled, the frequency shift of the carrier wave is doubled. If the frequency of the modulating signal is doubled, the time period during which the frequency shift occurs will be halved. Obviously, both amplitude and frequency changes of a complex voice signal will be extensive and it is not possible to

DIFFERENCE BETWEEN AM & FM		
Modulating Signal	Characteristic of Modulated Wave	
	Amplitude Modulation	Frequency Modulation
Amplitude	Change of carrier amplitude	Change of carrier frequency
Frequency	Rate of change of carrier amplitude	Rate of change of carrier frequency

Fig. 14.4   Difference between AM and FM

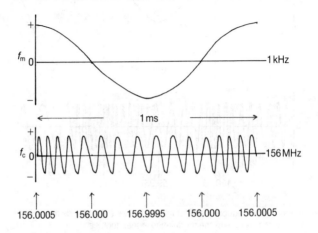

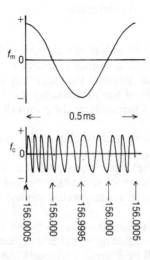

Fig. 14.5   Rate of change of the carrier frequency in an FM system

represent graphically such a wave. As with AM, FM bandwidth is represented by using frequency spectrum diagrams.

Terms employed in FM
(a) Frequency swing. This represents the difference between the minimum and maximum values of the instantaneous frequency of the FM wave.
(b) Frequency deviation (f_d). This signifies the peak difference between the instantaneous frequency of the FM wave and the carrier frequency during one cycle of modulation. Frequency deviation is directly proportional to the amplitude of the modulating signal.

As a consequence, the amplitude of the audio signal to be applied to the modulator in an FM transmitter is amplitude limited. This factor prevents the transmission signal bandwidth per channel from exceeding the rated maximum and avoids interchannel interference.
(c) Modulation index (m_f). The ratio of frequency deviation to the modulating frequency (f_m) i.e. $m_f = f_d/f_m$.

The three terms above refer to the FM signal whereas the following terms refer to the FM system.
(d) Rated system deviation ($f_{d.max}$). The maximum allowable frequency deviation in a given system. This is usually 75 kHz for broadcast FM stations and 5 kHz for maritime VHF FM services.
(e) Maximum modulating frequency ($f_{m.max}$). The maximum permitted modulating frequency within the audio baseband. For maritime VHF FM this is 2700 Hz.
(f) Deviation ratio (D). Ratio of the rated system deviation to the rated maximum modulating frequency of a system: 1.85 for maritime VHF FM systems.

Bandwidth of an FM signal
A standard formula for the determination of the bandwidth occupied by an FM signal is:

$$\text{Bandwidth} = 2\,(f_d + f_{m.max})\ \text{Hz}$$

Consider maritime VHF FM systems where the rated system deviation is 5 kHz and the maximum modulating frequency is accepted as 3 kHz. The bandwidth, on transmission, occupied by this signal is:

$$\text{Bandwidth} = 2\,(5 + 2.7)\ \text{kHz} = 15.4\ \text{kHz}$$

The same signal transmitted using the AM SSB J3E mode would possess a bandwidth of 2.7 kHz. However, signal quality is vastly improved when using FM, and noise is more readily counteracted. The greater signal bandwidth demanded by an FM signal means that frequency modulation is only used on the VHF band and bands above.

Phase modulation
A phase modulated carrier wave has its instantaneous phase varied in accordance with the characteristics of the modulating signal. The extent of the phase deviation is proportional to the amplitude of the modulating signal whilst the rate at which the phase is deviated is equal to the modulating frequency.

There are many types of phase modulation available but the most popular is the generic method of data or digital modulation called phase-shift keying (PSK).

Where the data bit stream to the modulator input possesses only two possible values (0 or 1) the system is termed binary and the modulation mode becomes binary-phase-shift keying (BPSK). If there are more than two possible bit combinations 00, 01, 10, 11

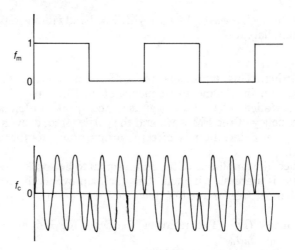

The phase of the carrier frequency is reversed at each logic
transition of the modulating signal. Both the amplitude
and frequency of the carrier signal remain unchanged.

(a)

Fig. 14.6(a)   Phase modulation by a square wave

for instance—the modulation mode becomes quadrature-phase-shift keying (QPSK) or
its variation called offset QPSK (O-QPSK). In BPSK the modulating signal changes the
phase of the carrier frequency at each data bit transition, 1 to 0 or 0 to 1 as shown in
figure 14.6(a)

In this system the frequency of the carrier wave remains constant whilst the phase is
changed, in this example, by 180° with each bit transition causing a phase change of the
modulated signal.

BPSK modulation is used in the terrestrial telex system, whilst BPSK, QPSK and
O-QPSK modes are all used for data modulation in the satellite segment of the GMDSS
radionet.

AM mode selection by an operator

In practice, the modulation mode may not be changed by the operator of the equipment
and has been determined by international and government regulatory bodies. The only
change which may be under operator control is the choice of the three modes of SSB
AM on a terrestrial MF/HF transceiver. An exception to this rule is when 2182 kHz is
selected for distress operation when the transceiver will automatically (if not correctly
selected by the operator) transmit the H3E mode.

This choice of mode—H3E, R3E or J3E—to be used depends upon the factors which
affect the communications link. These factors include the level of noise encountered,
which affects the S/N ratio and the quality of the receiving equipment. The use of J3E
greatly improves the S/N ratio by virtue of reduced bandwidth and more power
contained in the sideband to overcome the constant noise level. However, the use of the
J3E mode demands the use of a vastly superior demodulation system, to that needed for
H3E, in the receiver.

When establishing a communications link on MF/HF it is usual to commence calling
using the H3E mode which will be detected by the receiver even if the operator has

selected J3E. The converse will not be appropriate. If the transmission is made using J3E and the receiver is switched to H3E the call may not be detected.

To enable an AM signal to be demodulated in the receiver, both the sideband and the carrier must be present with sufficient amplitude to drive the signal detector circuit. When the H3E mode of modulation is used, both the transmitted carrier and the sideband contain power and both are applied to the 'envelope' detector circuit in the receiver. If the J3E mode of modulation is used, power will only be contained in the transmitted sideband, because no carrier exists. A receiver must therefore provide the carrier as a locally generated signal from the carrier re-insertion circuitry, which is automatically selected by the operator when switching to J3E. The locally generated carrier frequency and the sideband are then applied to the 'product' detector to produce the audio signal.

14.3   Telex communications

The main disadvantage of using telex for communications is that skilled keyboard operators are required. One of the main advantages is that a hard copy of all communications exists at both ends of the link. Telex communication uses digital signals on the link with a resulting increase in the transmission bandwidth. However, the data rate of transmission is slow at 50 bauds, enabling the system to use channels with a 3 kHz bandwidth similar to amplitude modulated voice channels, thus eliminating the need for increased bandwidth channels.

Terrestrial telex specifications are those produced by the CCIR as recommendation 476–4 'Direct-Printing Telegraph Equipment in the Maritime Mobile Service'.

A telex link uses teleprinters (which are essentially electronic typewriters) or, increasingly, computer terminals. Messages are formatted in the word processing section of the equipment and are then transferred to the 'line' section for onward transmission. Teleprinters use the CCITT ITA No.2 code (International Telegraph Alphabet Code) for communications. Each keyboard character is transposed to a five-element code of constant duration. This code is only able to provide 32 combinations but the use of the 'shift' key enables a further 26 combinations to be available.

All data communications systems encompass some form of error checking arrangement. Telex over radio (TOR) uses a seven-unit code. Each of the five-unit ITA characters is converted to seven-units as shown below in figure 14.6(b).

International terminology designates mark (logic 1) elements 'Z' and space (logic 0) elements 'A'. The seven-unit telex code derived from ITA code No.2 is described as 3A/4Z. All correct characters received must possess the 3/4 ratio and, consequently, a 3/4 parity checking system can be used to detect errors.

The two methods of error checking and correction used are: forward error correcting (FEC), CCIR Mode B, and automatic repeat on request (ARQ), CCIR Mode A. Both methods utilize the seven-element code and 3/4 parity checking.

Forward error correcting (FEC) system

Forward error correction is a procedure which permits the detection of transmission errors in the transmission link. As with all error detection methods, FEC relies upon subdividing the information into frames and attaching additional data bits to each one to provide some form of 'parity'. Each frame, including parity information, is checked on transmission and again on reception. If an error in the checksum exists the receiving station requests a retransmission of the error frame.

Coastal radiostations broadcast weather forecasts etc, using FEC encoded telex on

	Lower case	Upper case	ITA Code 2	7 unit code
		TELEX OVER RADIO 7-UNIT CODES		
1	A	–	ZZAAA	ZZZAAAZ
2	B	?	ZAAZZ	AZAAZZZ
3	C	:	AZZZA	ZAZZZAA
4	D		ZAAZA	ZZAAZAZ
5	E	3	ZAAAA	AZZAZAZ
6	F	%	ZAZZA	ZZAZZAA
7	G	@	AZAZZ	ZAZAZZA
8	H	£	AAZAZ	ZAAZAZZ
9	I	8	AZZAA	ZAZZAAZ
10	J	BELL	ZZAZA	ZZZAZAA
11	K	(	ZZZAZ	AZZZZAA
12	L	)	AZAAZ	ZAZAAZZ
13	M	.	AAZZZ	ZAAZZZA
14	N	,	AAZZA	ZAAZZAZ
15	O	9	AAAZZ	ZAAAZZZ
16	P	0	AZZAZ	ZAZZAZA
17	Q	1	ZZZAZ	AZZZAZA
18	R	4	AZAZA	ZAZAZAZ
19	S	'	ZAZAA	ZZAZAAZ
20	T	5	AAAAZ	AAZAZZZ
21	U	7	ZZZAA	AZZZAAZ
22	V	=	AZZZZ	AAZZZZA
23	W	2	ZZAAZ	ZZZAAZA
24	X	/	ZAZZZ	AZAZZZA
25	Y	6	ZAZAZ	ZZAZAZA
26	Z	+	ZAAAZ	ZZAAAZZ
27	Carriage return		AAAZA	AAAZZZ
28	Line feed		AZAAA	AAZZAZZ
29	Letter shift		ZZZZZ	AZAZZAZ
30	Figure shift		ZZAZZ	AZZAZZA
31	Space		AAZAA	AAZZZAZ
32	Unperforated tape		AAAAA	AZAZAZZ
33	Idle signal			ZZAAZZA
34	Idle signal			ZZZZAAA
35	Signal repetition ARQ			AZZAAZZ

Fig. 14.6(b)   Telex over radio seven-unit codes

frequencies in the MF and HF bands, to be received without acknowledgement by a vessel.

In the terrestrial FEC system each character is transmitted twice, the first (direct) transmission known as DX and the repeat transmission as RX. Upon receipt, both DX and RX characters are checked for parity. If either the DX or the RX character is correct it is printed. If neither the DX or the RX character is correct it is rejected and a space is printed. Signal fading is a major problem particularly when using frequencies within the HF bands for global communications. To reduce the effects which fading has on the telex signal, the RX transmission is time delayed after the DX transmission. The time delay corresponds to four characters between the specific DX and RX transmitted character. This allows for time diversity reception, to overcome fading problems, of up to 280 ms.

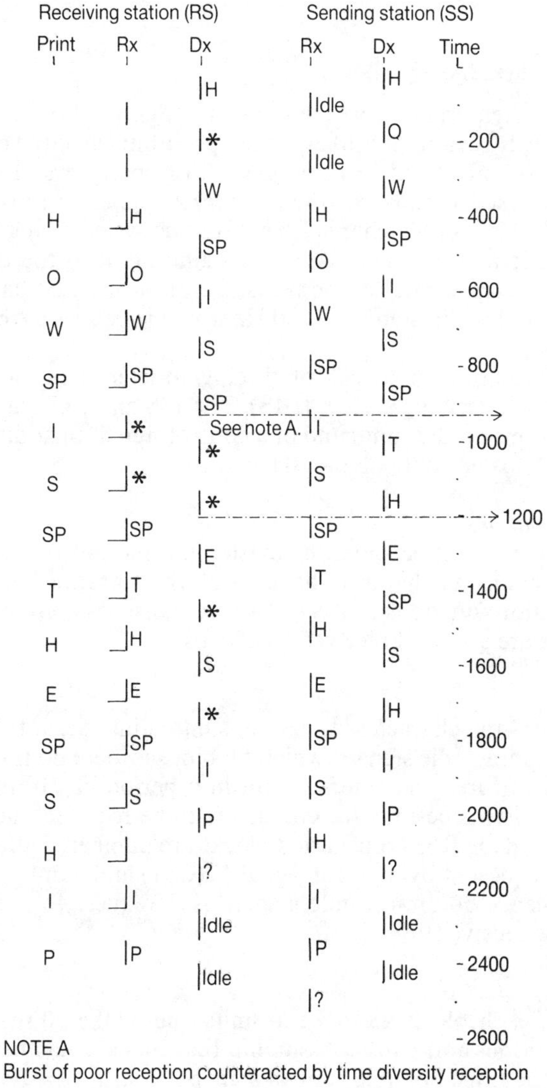

| Receiving station (RS) | | | Sending station (SS) | | |
Print	Rx	Dx	Rx	Dx	Time
		H	Idle	H	
|		*	Idle	O	– 200
|		W	H	W	– 400
H	H	SP	O	SP	
O	O	I	W	I	– 600
W	W	S	SP	S	
SP	SP	SP		SP	– 800
I	*	See note A	I	T	– 1000
S	*	*	S	H	
SP	SP	*	SP		→ 1200
		E	T	E	
T	T	*	H	SP	– 1400
H	H	S	E	S	– 1600
E	E	*		H	
SP	SP	I	SP	I	– 1800
S	S	P	S	P	– 2000
H		?	H	?	
I	I	Idle	I	Idle	– 2200
P	P	Idle	P	Idle	– 2400
			?		– 2600

NOTE A
Burst of poor reception counteracted by time diversity reception

Fig. 14.7 Radiotelex FEC mode of operation

Each message transmission commences with the sending of phasing signals on both the DX and RX streams. The 'end of transmission' signal consists of three consecutive 'idle signals' transmitted in the DX stream only, immediately after the last traffic character in the DX stream.

Selective FEC

SEL-FEC is used in the shore-to-ship direction where a coastal radiostation is able to send broadcast messages to a specific vessel by using its selective call number in conjunction with broadcast telex. It is also possible to use the 'store-and-forward' mode in SEL-FEC where telex traffic, from a subscriber ashore, is stored and then automatically sent to ships at a convenient time even when the vessel is in port and not able to use its transmitter.

Automatic repeat on request (ARQ) system

When using the FEC system in areas where there is a high level of interference or the signal suffers from a low S/N ratio, it is likely that a prohibitively large number of blank spaces will appear on the print-out in response to lost characters. To overcome this problem ARQ may be used because it offers a greatly superior error detection and correction process. ARQ is a single-channel synchronous system using the same seven-unit error detecting code as FEC. The baseband signal input to the terminal is at 50 bauds whereas the modulation rate on the transmission link is 100 bauds. The centre frequency of the baseband audio signal is 1700 Hz which produces a frequency shift on transmission of 170 Hz.

ARQ transmits information in blocks of three characters from a pre-determined sending station (ISS) to a receiving station (IRS). The ISS and IRS stations are able to interchange their functions on the command of a control signal. In addition, one station will be designated the 'master' and one the 'slave'.

Master and slave designations
The station initiating the circuit becomes the master and the called station is the slave. This configuration remains unchanged throughout the duration of the radio link regardless of which station functions as ISS or IRS. A transmission timing cycle, during which three characters are sent and received, is 450 ms.

The sending station
A sending station groups the characters to be transmitted into sets of three, producing 21 signal elements, including 'idle signals' which fill blocks where no traffic information is to be sent. A block of three characters is sent in a period of 210 ms after which a transmission pause of 240 ms permits the characters to be received and acknowledged by the control signal from the RS. Transmitted blocks are numbered alternately, block 1 and block 2, and are requested by 'control signal 1' (CS1) and 'control signal 2' (CS2) respectively. The sequence continues unless there is a request for repetition due to mutilated signals being received.

The receiving station
After the reception of each block the RS transmits one of the 70 ms control signals initiating a 380 ms transmission pause. Assuming that block 2 has been received, the IRS transmits one of the three possible control signals:
CS1 : if the block has been correctly received,

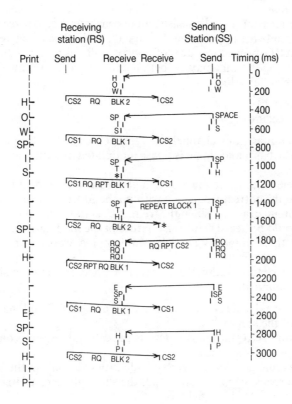

Fig. 14.8   Radiotelex ARQ mode of operation

CS2 : to request a repeat of a mutilated block 2,
CS3 : at the end of a cycle to request change-over.

Change-over
The IRS transmits its control signal CS3 to request a change-over after which both stations reverse their designations. CS3 is transmitted in response to:

• a received block containing no information (three β),
• a received block containing the characters 'shift', + and ?

In response to the CS3 signal, the ISS transmits a block containing the characters, idle signal β, idle signal α and idle signal β, after which the IRS becomes the ISS and vice versa. The sequence now operates in the reverse direction.

The change-over system is also effected if the 'Who Are You? (WRU) request is received by the IRS.

End-of-communication

The end-of-communication signal consists of the character α sent three times by the ISS. The IRS now transmits the appropriate CS signal and both stations revert to standby.

The FEC and ARQ procedures are, of course, fully automatic. From an operator point of view, telex operations may be better understood by following the procedure required to establish a telex link.

Telex operating procedures

All terrestrial communications channels carrying telex use two frequencies which may be paired or not. The baseband signal is phase modulated onto the carrier frequency and designated F1B.

To establish a telex link using one of the high frequency bands it is necessary for the operator to choose the correct frequency within a band from the list made available by the shore-based administration. When the frequency is tuned-in on the mobile receiver a 'warbling tone' should be heard. This is a mixture of the modulation methods A1A/F1B and signifies that the channel is available and that a watch is being kept on that frequency.

The following procedure outlines the operations needed between ship and shore to establish telex communications.

- The ship operator calls the coastal radio station using ARQ and the selective identity (call number) of the station.
- The coast station's ARQ equipment detects the call and responds on the corresponding (paired) transmit frequency.
- Both stations exchange telex answerback identities to confirm the link.
- Once the answerback signal is confirmed, the ship will receive either:
    (i)   NO TRAFFIC HELD/QRU, or
       (ii)   GA+?
  GA+? is the invitation to transmit.
- The ship now requests one of the two modes of operation available:
    (i)   DIRTLXxy+ for a direct connection to be made to the subscriber in a conversation mode, or
    (ii)   TLXxy+ where 'store-and-forward' mode is to be used. In this mode the traffic is stored by the coast station until a later time.
  Where 'x' is the telex destination country code and 'y' is the desired telex subscriber number.
- After the message has been sent the 'end-of-facility' code KKKK is transmitted.
- The coast station now transmits 'housekeeping' information in the form of date, time and call duration.
- If no further communication is required, the code BRK+ is used to clear the radio circuit.

Assuming a call is to be made to a telex subscriber in Bonn, Germany, number 987 6543, via the UK HF radio station Portisheadradio, the following print-out is an example of the information displayed on the NBDP.

GA+

DIRTLX0419876543+
MOM
Answerback exchange
MSG+?

** Exchange message text via keyboard.**

KKKK
date.....and time group. (1212310893)
durationofcall. (2.4 mins)
GA+?
BRK+

A variety of telex command codes and abbreviations may be used in a telex circuit as shown in operator manuals.

14.4   Data over radio

The word data refers to groups or strings of digital bits which carry information. Data over radio (DOR) is the term used to describe the sending of written text over a radio link. Traditionally, the transfer of information over a radio communications link has been achieved by the use of analogue modulation methods. Increasingly, digital modulation is replacing analogue modulation for a variety of reasons. As with most principles of electronic communications engineering the choice of a system is the result of a compromise. In the case of digital communications the compromise is between channel access time and signal bandwidth.

Unmodified digital transmission uses far greater bandwidth for the transfer of information than most analogue transmissions. However, because digital transmission may be compressed, a smaller bandwidth can be achieved, thus maintaining high channel capacity on the limited frequency spectrum. This fact coupled with the numerous advantages of digital transmission means that, within the next decade, digital modulation will totally replace analogue modulation in communications systems. The main advantages of digital communications are as follows:

- information may be communicated at high speeds leading to a very efficient use of channel time,
- error detection and correction methods may be used to verify the message,
- the communications link may be fully automated, and
- computers may be used on the link, thus expanding the data capabilities.

The terrestrial part of the GMDSS radionet uses digital communications for the transmission of radiotelex and NAVTEX, both of which are known as Narrow Band Direct Printing (NBDP) systems. Both systems operate using the CCIR recommendation 476–4, which fully stipulates the principles of 'Direct-Printing Telegraph Equipment in the Maritime Mobile Service'. As its name suggests, DSC also uses digital methods of communications as stipulated in the CCIR recommendation 493–3 'Digital-Selective-Calling System for Use in the Maritime Mobile Service'.

14.5   Transceiver systems

Modern transceiver systems are greatly simplified, when compared with traditional systems, in order to permit their use by non-technical personnel. Whilst simplification has led to ease of operation it has in no way reduced the equipment capabilities. In fact, modern transceivers are far more versatile than their older counterparts.

A transceiver, either for use on MF, HF or VHF, is under the control of a microprocessor with all operator commands input via a keypad. Whilst each manufacturer produces equipment to a different design the operator controls essentially provide the same function. The use of a microprocessor for system control enables many of the functions to be automated, which again leads to simplification. In addition, the central processor will not permit the operator to use the wrong modulation system on a selected channel, or communicate using 'cross-band' or 'out-of-band' frequencies.

Channel frequency selection by the operator is via the keypad, which commands computer software to tune the transmitter and receiver to the correct paired frequencies, to select the correct modulation system and the correct peripherals. If, as happens regularly following a WARC conference, the entire global HF channel frequencies are redesignated, a suitable change of EPROM will rapidly re-align the system.

The maritime MF/HF system uses amplitude modulation (AM) for modulating the baseband audio signal onto the carrier frequency. ITU radio regulations prohibit the use of double sideband amplitude modulation (DSB). All voice communications on the MF/HF bands must use single sideband (SSB) modulation. DSB (designated A3E) may only be used in order to receive DSB signals such as weather forecasts from local broadcast commercial stations.

There are three modes of SSB modulation used on transmission. They are described below including their internationally designated notations:

- **SSB H3E**—a single sideband containing the signal intelligence plus a carrier frequency of maximum amplitude.
- **SSB R3E**—a single sideband as above plus a carrier frequency of reduced amplitude.
- **SSB J3E**—a single sideband as above plus a fully suppressed carrier frequency amplitude. In fact, the carrier frequency amplitude is suppressed to the point where it can be considered not to exist.

The use of H3E should be avoided because the actual radiated power contained in the sideband (the intelligence signal) may be as low as 12 W even when the transmitter is switched to full power and radiating 200 W PEP. In this example, 188 W is contained in the carrier and does not enhance the intelligible link.

Additionally, the high-power carrier frequency is transmitted continuously even when no modulation exists. This means that the system, when compared to using J3E, is very inefficient.

R3E provides a much better ratio of sideband to carrier power where the same 200 W PEP may contain 100 W in the sideband and 100 W in the carrier.

J3E is a better choice whereby no power exists in the carrier and virtually all of the 200 W PEP is contained in the sideband.

Basically, the choice of AM mode depends upon the quality of the communications link and the receiving equipment demodulation stage. Noise is always the major problem in a radio communications link. The larger the signal amplitude of the sideband on transmission, the better will be the overall S/N ratio of the link. Consequently, J3E should be used as often as possible. However, J3E can only be used if the receiver demodulator is equipped to handle a signal with no carrier. If there is doubt R3E should be used. Section 14.2 provides details of AM and the conditions under which the three modes would be selected.

The essential task of tuning the antenna to resonance with the selected frequency at the transmitter output is now done automatically eliminating the need for a skilled operator—to match, load and tune for a 'dip'—as the traditional communications officer did. However, if the automatic tuning system fails it is also possible to tune the antenna unit manually on a modern GMDSS compliant MF/HF transceiver. Fully automating the system in this way enables the equipment to be controlled from the navigation position, thus eliminating the need to carry a dedicated radiocommunications officer. Indeed, as the GMDSS regulations state 'It shall be possible to initiate distress alerts from the position from which the ship is normally navigated'.

A modern MF/HF terrestrial communications transceiver system, when fitted with a personal computer (PC) with NBDP, provides facilities for voice, radiotelex, data communications and DSC, and fully complies with the GMDSS regulations for terrestrial MF/HF communications.

Frequency synthesis

It was necessary on traditional transceiver systems to use a separate quartz crystal to generate the carrier frequency for each channel to be used. This is clearly an undesirable situation on a vessel which is on an international voyage and needs to communicate with dozens of countries using any one of the many hundreds of channels available. To alleviate this problem modern transceivers use frequency synthesis to generate the multitude of frequency channels required.

The concept of frequency synthesis has been known since the 1940s but it is only in the last decade that, due to the development of integrated digital circuit technology, large-scale synthesis has become a practicality. Frequency generation within the system relies for its operation on the phase locked loop (PLL) which comprises a phase comparator, a low-pass filter, a voltage-controlled oscillator (VCO) and a programmable divider.

There are two inputs to the phase comparator, one from a highly stable reference oscillator and the other from the output of the frequency divider. The output of the phase comparator is the 'error voltage' which is proportional to the phase difference of the two input signals. The error voltage is filtered to produce a d.c. which is applied to the VCO to control its output frequency. This, in turn, is fed back to the programmable divider ultimately to control the second input to the phase comparator. Basically, if the two inputs to the phase comparator are of the same phase the loop is considered to be in a stable state and the output frequency f_O will be correct. If a discrepancy exists in the phase difference between the two inputs the error voltage will cause the VCO to re-tune to the correct frequency when the loop becomes stable again. It is so arranged that the d.c. applied to the VCO will increase or decrease depending upon whether the phase error is advanced or retarded on the zero phase point. The up/down

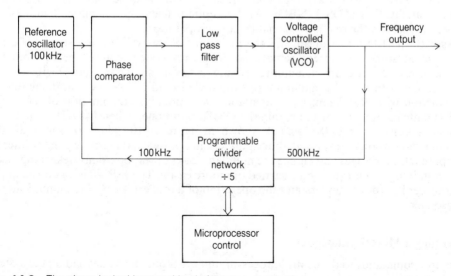

Fig. 14.9   The phase locked loop and basic frequency synthesis

change of p.d. to the VCO causes the oscillator frequency to increase or decrease accordingly.

If the programmable divider is capable of dividing in integers from $N = 1$ to 10 and the reference frequency is 100 kHz the following output frequencies could be produced:

f_R	N	f_O
100 kHz	1	100 kHz
100 kHz	2	200 kHz
100 kHz	3	300 kHz
100 kHz	4	400 kHz
100 kHz	5	500 kHz
etc		

Using this very simple PLL it is possible to produce ten highly stable output frequencies from one single crystal oscillator. Obviously, far more complex systems may be used to produce the many hundreds of frequencies required in a modern transceiver.

It is likely that the programmable divider would be controlled by the system micro-processor and that all the frequency generation would be under the command of software. Software may be more readily changed than hardware when communication channels have been re-aligned by committee at a WARC meeting.

14.6   The MF/HF transceiver

A large number of manufacturers worldwide produce MF/HF transceivers for use within the terrestrial component of the GMDSS radionet. Modern MF/HF transceivers are much more complex than those which were used by the traditional radio officer. The massive advances which have taken place in communications technology design have led to transceivers which are very user friendly, highly reliable and indistinguishable from a satellite BDE. DSC, NBDP, MSI reception, telex and telephone facilities are fully integrated in the one unit to enable ease of operation.

As with the BDE part of an SES the main unit of the MF/HF transceiver is often mounted at some convenient remote position with system control being from the wheelhouse or chartroom via a controller similar to that shown in figure 14.10.

JRC, one of the biggest manufacturers of GMDSS radio equipment in the world, in conjunction with Raytheon, make available every possible combination of terrestrial and satellite equipment for use within the GMDSS radionet. Their MF/HF transceiver system is based on the JSS710/720, shown in figure 14.10, which performs all the communications functions listed above plus has an optional data terminal and printer to compose telex messages and enable the mundane tasks of shipboard housekeeping, such as maintaining inventories, to be carried out more easily. The JSS720 is a good example to consider in order to understand the operation of a modern MF/HF communications transceiver.

Operating a MF/HF transceiver

Operator commands to the central processor unit are input via the keypad on the system main controller.

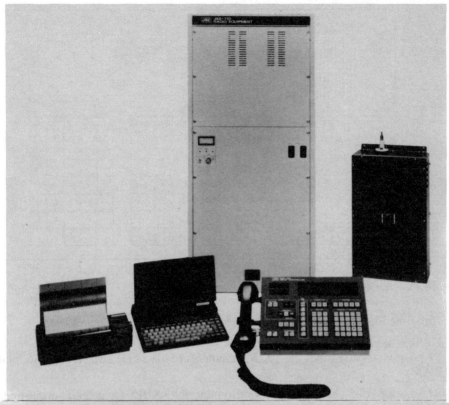

Fig. 14.10   The JRC JSS710/720 MF/HF transceiver system

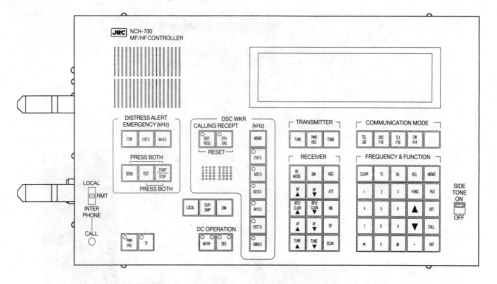

Fig. 14.11 The JSS710/720 controller

Control functions are sub-divided into areas of control as listed below.

- **Distress alert keys**

[2182] Automatically tunes the TX/RX channels to 2182 kHz and sets the modulation H3E AM.

[2187.5] or [8414.5] Select DSC operation and set TX/RX frequencies accordingly.

[SEND] & [START] Keys pressed together when [H3E] or [TEL] is elected, transmit the two-tone alarm signal (1300 Hz and 2200 Hz). When [DSC] is selected a DSC call is made.

[TEST] & [START] Keys together test the two-tone alarm signal through the loudspeaker under the same conditions as above. The alarm signal is not transmitted.

- **DSC WKR (watch-keeping receiver)**

[2187.5 through to 16804.5] These six keys select the frequencies to be scanned for DSC channels watchkeeping.

[MEMO] Used to set DSC frequencies to be scanned.

[DISTRESS or OTHERS] Reset the receiver after a call has been received. The LED on the key's lights to indicate signal priority.

- **Transmitter keys**

[TUNE] Pressed to activate automatic TX tuning system.

[PWR RDC] Each press sequentially changes the TX output power. FULL – MEDIUM or LOW.

[TONE] Initiates a TX testing tone.

- **Receiver keys**

[RX MODE] Each press sequentially selects the receiving mode. J3E – USB telephony, F1B – FSK DSC and telex, A1A – CW Morse telegraphy (until February 1999) and H3E – AM broadcast stations.

[BW] Selects the receiver bandwidth. BW – 0.3k (kHz) for DSC, BW – 1k for Morse, BW – 3k for SSB telephony, BW – 6k or 12k for broadcast stations. Weather reports etc.

[AGC] Selects automatic gain control level. AGC – FAST for DSC, AGC – SLOW for telephony and AGC – OFF for Morse.

[ATT] Sets the attenuator on/off to change the receiver amplification. Should be off when changing mode or frequency.

[NB] Noise blanker on/off switch. When on, the NB reduces noise.

[SP] Turns speaker on/off.

[SCAN] Activates scanning of selected frequencies.

[UP] [DOWN] keys are used to increase or decrease the levels of the system shown.

[RF] Used to increase or decrease the receiver amplifier gain. Shown as an eight-stage bar graph on the LED display.

[BFO/CLARI] Used to clarify or finely adjust an input to the demodulator. Effectively finely tunes the received signal.

[AF] Changes the level of the volume from the speaker.

[TUNE] Tunes the receiver frequency along the selected band in 10 Hz steps.

- **Frequency and function keys**
[FILE] [CALL] [OFF] [*] [#] Used for DSC operation. File enables specific messages to be held in memory.

[TX] [RX] Sets TX or RX for frequency selection.

[RCL] Recalls from memory the frequencies of the ITU channels.

[MEMO] Used to write a TX/RX frequency, or position data for DSC, into memory.

[FUNC] Special functions select including memory channel allocation and scanning duration.

[UP] [DOWN] Up and down arrow keys to select a TX/RX channel. Also used in the DSC mode for various items.

[ENT] Command enter key.

- **Communications mode keys**
[TEL J3E] SSB Radio-telephony mode.

[DSC F1B] DSC communications mode.

[TLX F1B] Radiotelex.

[CW A1A] Morse radio-telegraphy. (Until February 1999)
- **Other keys**
[LOCAL] Overrides remote controllers giving priority to the one selected.

[DUP/SIMP] Duplex/simplex channel selection.

[DIM] Dimmer control.

[MF/HF] [SES] Selects terrestrial or satellite units for d.c. operation.

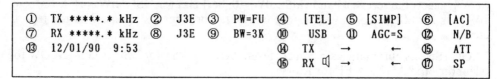

Fig. 14.12   The LCD display. (Courtesy JRC/Raytheon)

☆☆☆   SSB Telephony   ☆☆☆

Follow these steps:

①   Setting the TX and RX frequencies (setting different frequencies):

Example)   Setting TX frequency at 4180 kHz and RX frequency at 4400 kHz

a)   You set the TX frequency first:
Press the [TX] key on the FREQUENCY & FUNCTION key cluster,
and the following screen appears.

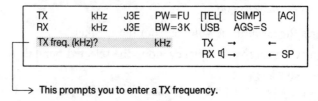

→ This prompts you to enter a TX frequency.

Using numerical keys, type [4] [1] [8] [0].
Press the [ENT] key for confirmation.

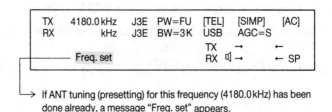

→ If ANT tuning (presetting) for this frequency (4180.0 kHz) has been
done already, a message "Freq. set" appears.
(It disappears in a few seconds.)

Fig. 14.13   Setting-up the equipment for SSB telephony use. (Courtesy JRC/Raytheon)

A large liquid crystal display (LCD) is used to display all the communications parameters as shown in figure 14.12.

The reader should have no difficulty in interpreting the display. The time is 0953 on 12
January 1990. No TX/RX frequencies have yet been selected. However, the modulation
method is J3E, the selected bandwidth is 3 kHz, the upper sideband is selected, the AGC
is slow, the simplex mode is displayed and indeed [TEL] is indicating that SSB
radio-telephony has been selected. The transmitter power is switched to full. Other
indications show that the noise blanker is on, the attenuator is on and the speaker is on.
Lastly, the TX (14) and RX (16) displays are not yet showing a bar graph.

It is a simple matter to set-up the equipment, for instance, for SSB telephony
operation. When selecting SSB and pressing the TX key, the LCD will provide the
information shown in figure 14.13.

* If not, a message "Tuner no-setting" is displayed. Press the [Tune] key
on the TRANSMITTER key cluster. Automatic tuning is performed, and
a message "Tuning ok" appears in several to several tens of seconds.

b) Then, set the RX frequency in the following manner:

Press the [RX] key on the FREQUENCY & FUNCTION key cluster, and
the following message appears.

```
┌──────────────────────────────────────────────────────────────┐
│  TX   4180.0 kHz   J3E   PW=FU   [TEL]   [SIMP]   [AC]          │
│  RX           kHz   J3E   BW=3K   USB    AGC=S                  │
│                                   TX      →        ←           │
│  RX freq. (kHz)?            kHz   RX  ◁→         ← SP           │
└──────────────────────────────────────────────────────────────┘
```

→ This prompts you to enter a RX frequency.

Using numerical keys, type [4] [4] [0] [0].
Press the [ENT] key for confirmation.

```
┌──────────────────────────────────────────────────────────────┐
│  TX   4180.0 kHz   J3E   PW=FU   [TEL]   [SIMP]   [AC]          │
│  RX   4400.0 kHz   J3E   BW=3K   USB    AGC=S                  │
│                                   TX      →        ←           │
│                                   RX  ◁→         ← SP           │
└──────────────────────────────────────────────────────────────┘
```

→ No message is displayed if you have set the RX frequency.

② Setting the communication mode:

Press the [TEL J3E] key on the COMMUNICATION MODE cluster.

```
┌──────────────────────────────────────────────────────────────┐
│  TX   4180.0 kHz   J3E   PW=FU   [TEL]   [SIMP]   [AC]          │
│  RX   4400.0 kHz   J3E   BW=1K   USB    AGC=S                  │
│                                   TX      →        ←           │
│                                   RX  ◁→         ← SP           │
└──────────────────────────────────────────────────────────────┘
```

Upon this, modes indicated by the ▓▓ are set automatically.

Fig. 14.13 (cont.)

Using DSC for communications

If the DSC unit is required for immediate distress alerting. The [SEND] & [START]
keys are pressed simultaneously causing the following message (figure 14.14) to be
transmitted five times. The ship's position and UTC are interfaced directly from the
satellite navigation equipment.

If, however, there is time to input other commands a full distress call and message can
be sent using information previously stored in memory.

Other calls using the DSC equipment are shown on the flowchart of DSC operation in
figure 14.15.

③ Setting the receiving sound volume:

a)   Press the [RF▲] or [RF ▼] key, and the previously set RF gain is
displayed. It disappears in about two seconds.

```
TX    4180.0 kHz   A1A   PW=FU   [CW]   [SIMP]      [AC]
RX    4400.0 kHz   A1A   BW=1 K  CW     AGC=S
                                 TX       →       ←
                RF → ||||||||||| ←   RX  ◁ →       ← SP
```

b)   Set the AF gain at a suitable value using the (AF ▲] or [AF ▼] key.

```
TX    4180.0 kHz   A1A   PW=FU   [CW]   [SIMP]      [AC]
RX    4400.0 kHz   A1A   BW=1 K  CW     AGC=S
                                 TX       →       ←
                                 RX  ◁ → |||||||   ← SP
```

④ If you have caught a desired signal, adjust the sound volume using the
[BFO/CLARI ▲] or [BFO/CLARI ▼] key. The entry disappears in a few seconds.

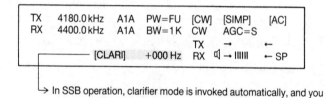

```
TX    4180.0 kHz   A1A   PW=FU   [CW]   [SIMP]      [AC]
RX    4400.0 kHz   A1A   BW=1 K  CW     AGC=S
                                 TX       →       ←
          [CLARI]    +000 Hz    RX  ◁ → |||||||    ← SP
```

↳ In SSB operation, clarifier mode is invoked automatically, and you
can set clarification within the range of ± 200 Hz in 1 Hz steps.

⑤ If you want to turn off the sound from the panel speaker, press the [SP] key.
The [SP] key is used to turn on and off the speaker. "SP" on the screen
disappears if you press the [SP] key.)

```
TX    4180.0 kHz   A1A   PW=FU   [CW]   [SIMP]      [AC]
RX    4400.0 kHz   A1A   BW=1 K  CW     AGC=S
                                 TX       →       ←
                                 RX  ◁ → |||||||   ←
```

⑥ When the above setting has been done, press the [TX] key on the panel to
turn on the high-voltage power. Then, perform communication using the
handset.

Fig. 14.13   (cont.)

System description

The chart in figure 14.16 shows the typical outline characteristics of a modern MF/HF
transceiver system. The reader should, at this stage, be able to interpret the chart with
little difficulty.

RAPID DISTRESS ALERTING USING DSC		
Operation	Response	Comment
Press SEND, and	The following message is printed and automatically transmitted 5 times.	
Press START simultaneously	This call attempt is transmitted again after a random delay of between 3–1/2 and 4–1/2 minutes from the start of the call.	
	Message; 05/15 12.34 CALL: MF/HF FORMAT: DISTRESS N of D: UNDES. DISTRESS POSITION: N54.27 WO12.45 DIST–UTC: 12.34 TELECOMM: J3E TEL EOS: EOS	Input position and time information drawn from MEMO programmed in advance.

Fig. 14.14   Rapid distress alerting using DSC. (Courtesy JRC/Raytheon)

An overall outline system description follows in order to explain how modern electronic circuit technology is providing the necessary equipment for communications.

It has previously been stated that modern communications equipment relies on frequency synthesis and phase locked loops in order to generate the huge number of channel frequencies required for operation throughout the world. The system diagram in figure 14.17 shows how the JRC JSS720 MF/HF transceiver achieves this.

Briefly, the transmitter system consists of three main circuit boards, the SSB generator, the frequency converter and the power amplifier unit. A baseband audio signal from the microphone is applied via the local control board to the COM circuit on the SSB generator board, where it undergoes amplitude processing and frequency limiting before being passed to the AF switch. Other inputs to this switch may be the audible two-tone alarm signal, NBDP, DSC signals or an audio frequency tone. Output from the AF SW, in the telephony mode, is the processed audio signal which is coupled to the LSB balanced modulator for up-converting to approximately 455 kHz. The J3E signal thus produced is band pass filtered and amplified before being coupled to the gain control circuit which sets the required signal amplitude. If H3E has been selected, a first carrier frequency of 455 kHz is applied to the signal line via an attenuator. A voice operated carrier (VOC) system is used in order to improve system efficiency. Briefly, the transmitter will not emit energy until a voice signal is present at the output of the SSB generator board.

The first carrier modulated signal at 455 kHz is applied to the first frequency mixer on the frequency converter board where it is upconverted to 70.455 MHz by the other fixed signal input to the mixer; 70 MHz (fL1). Frequency generation and stability is controlled by a 20 MHz master oscillator and an arrangement of phase locked loops. Transmitted channel frequencies are set at the output of the second mixer by the two

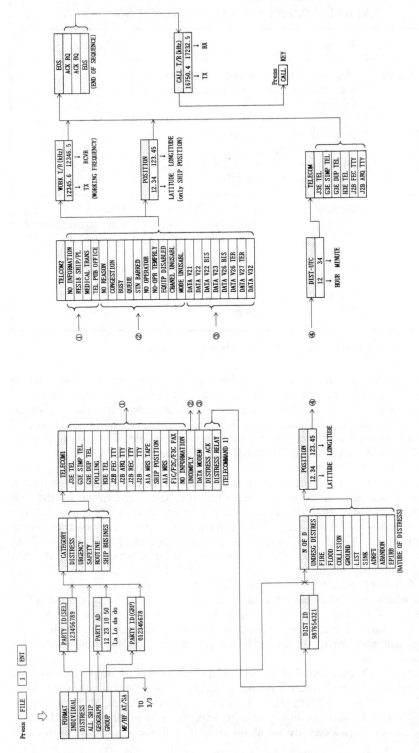

Fig. 14.15   Flow chart showing DSC calling operations. (Courtesy JRC/Raytheon)

OUTLINE MF/HF TRANSCEIVER SPECIFICATIONS	
Frequency band.	Transmit: 1.6 – 27.5 MHz variable in 100 Hz steps. Receive: 90 kHz – 29.9999 kHz variable in 10 Hz steps.
Frequency selection.	400 used channels. Set from memory or the keypad. 2182 kHz, 2187.5 kHz and 8414.5 kHz set by single action.
Modes of emission.	2182 kHz (H3E), USB (J3E), FSK (F1B/J2B) and CW (A1A).
TX Output power.	1.6 – 27.5 MHz. 400W PEP. 1.6 – 27.5 MHz. 60W PEP. (when using DC power supplies)
Power reduction.	Three steps.
TX signal bandwidth.	3 kHz for J3E/H3E. 0.5 kHz or less for F1B/J2B and A1A.
RX System.	Double superhet; 1st IF: 70. 45 kHz, 2nd IF: 455 kHz.
RX frequency band.	90 kHz – 29.999999 MHz.
Reception modes.	CW(A1A). MCW(A2A/H2A), DSB(A3E). USB/LSB(R3E/H3E/J3E). FSK(F1B/J2B). FAX(F3C).

Sensitivity.	90 – 1600 kHz: 10 μV or less (CW)
	                30 μV or less (DSB).
	1.6 – 30 MHz: 2 μV or less (CW).
	                6 μV or less (DSB).
	                3 μV or less (SSB).

Selectivity.	Bandwidth switchable: 12 kHz, 6 kHz, 3 kHz, 1 kHz and 0.3 kHz.
DSC Watchkeeping RX and transceiver.	
RX frequencies.	2187.5 kHz, 4207.5 kHz, 6312 kHz, 8414.5 kHz, 12577 kHz and 16804.5 kHz.
Frequency scanning.	Scans and/all of the above.
Mode.	F1B/J2B.
Sensitivity.	RF input level 1 μV. Symbol error rate 1×10^{-2} or less.
Protocol.	CCIR recommendation 493–3 & 541–2.
Mod/demodulation.	1,700 Hz $+/-$ 85 Hz.

Fig. 14.16 MF/HF transceiver outline specifications. (Courtesy JRC/Raytheon)

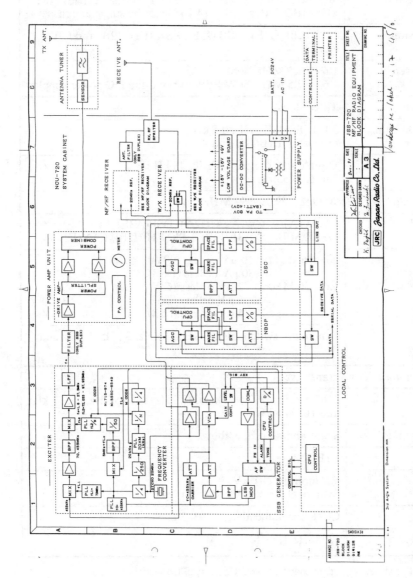

Fig. 14.17 JSS-720 HF/HF transceiver system diagram. (Courtesy JRC/Raytheon)

input signals. One, of 70.455 MHz, carrying the audio modulation and the other variable in the band between 70.055 MHz and 100.4549 MHz (fL2). Frequency fL2 is a high stability signal derived from the PLL fL2/N with the frequency dividing ratio under the command of a data line (N code) from the CPU control circuit on the SSB generator board. The channel frequency output from mixer two, in the band 1.6 to 27.5 MHz is filtered and applied to the power amplifier unit (PA).

Signal amplitude at the input to the PA is approximately 5 mW. Any transmitter PA stage must amplify all frequencies in the band equally and, consequently, it is called a linear amplifier. As this stage handles most of the power consumed by the transmitter it is essential that the efficiency of the stage is high. To achieve this a 'push/pull' arrangement of power transistors is used in a class-B (telephony signals) or class-C (telex/data signals) mode of operation. Consequently, the incoming signal is split, power amplified and recombined. The 5 mW signal is firstly amplified by the drive amplifier to approximately 15 to 30 W before the push/pull PA stage further amplifies it to 300 W for application to the antenna tuner and matching circuit.

Automatic antenna tuning is standard on modern MF/HF transceivers, although provision is made to tune manually the antenna output stage as required by the GMDSS regulations.

NBDP signals for telex and DSC working are generated on their respective circuit boards before being applied to the AF SW circuit on the SSB generator board where they are selected for transmission as commanded by the CPU control processor. The signals then follow the same processing path as previously described.

Signal processing in the receiver stages again relies upon locally generated frequencies produced by a master oscillator and a number of PLLs.

The standard 20 MHz frequency is taken from the oscillator on the frequency converter board, divided down to 5 MHz and applied to board LOOP 2. This board generates the fixed input to the second mixer circuit on the RF amplifier board. LOOP 1 circuitry generates the variable frequency, in the band 70.545 – 100.45499 MHz, first local oscillator input to the first mixer.

An input signal from the antenna is applied to the RF tune board where, after passing through receiver protection circuits, it is switched through the appropriate bandpass filter circuit. There are eight filter circuits because it is not possible to produce a filter with sufficient bandpass range and selective quality to cover the whole frequency range. The signal frequency is upconverted by the variable first local oscillator input to the first mixer to produce the fixed first intermediate frequency (IF) of 70.455 MHz. After amplification and filtering, the signal is down converted by the second mixer to the second IF of 455 kHz and applied to the IF filter board. Signal filtering on this board reduces noise and eliminates interfering signals. The filter block, comprising diode switches and four bandpass filters, is selected depending upon the mode of the received signal. As an example, an SSB telephony signal would be switched through the 3 kHz bandwidth filter.

A very large amount of signal amplification now occurs on the IF amplifier board, where the receiver automatic gain control (AGC) voltage is derived and demodulation takes place. An AF signal thus produced is applied to the control board for processing before being applied to the appropriate peripheral.

MF/HF transceiver self-testing

In the unlikely event of a fault occurring, it is possible using the information displayed by the LCD and a combination of keypad commands, to diagnose the area of a malfunction. Of course, the system may also be used to verify the performance of the equipment.

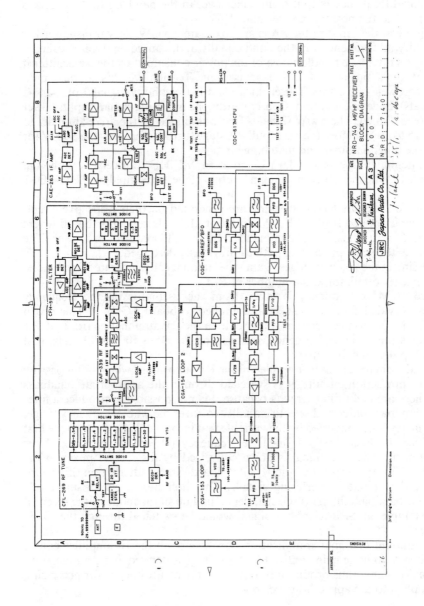

Fig. 14.18   JSS-740 receiver system diagram. (Courtesy JRC/Raytheon)

① Before self-test, the screen should look like.

```
TX   6200.0 kHz   A1A   PW=FU  [CW]    [SIMP]      [AC]
RX   6200.0 kHz   A1A   BW=1 K  CW     AGC=0
                                TX    →       ←
                                RX ◁ →       ← SP
```

② Press the [FUNC] key on the FREQUENCY & FUNCTION key cluster, and the
following screen appears.

```
– – – MENU P.1 – – –                [Item : ↑ ↓ +ENT]
→ * Self check
```

③ Press the [ENT] key on the FREQUENCY & FUNCTION key clusters to call up
the following screen.

```
– – – Self check – – –
1. RX                   2. WKR        3. DSC
4. EXCITER              5. TUNER      6. POWER AMP
```

④ To test the MF/HF receiver, press the numerical key [1] on the FREQUENCY &
FUNCTION key cluster.

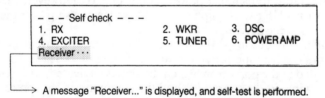

```
– – – Self check – – –
1. RX                   2. WKR        3. DSC
4. EXCITER              5. TUNER      6. POWER AMP
Receiver · · ·
```

→ A message "Receiver..." is displayed, and self-test is performed.

⑤ After self-test, a message "OK" is displayed if all the MF/HF receiver circuits
are normal.

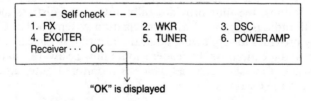

```
– – – Self check – – –
1. RX                   2. WKR        3. DSC
4. EXCITER              5. TUNER      6. POWER AMP
Receiver · · ·   OK
```

"OK" is displayed

* If there is any trouble with a circuit. A message indicating a defective
circuit is displayed.

Fig. 14.19   Self testing procedure for an MF/HF transceiver. (Courtesy JRC/Raytheon)

If, for example, there is a malfunction in the RX, a code number in the range ERROR 60 to 67 will appear on the LCD. Whichever error number is displayed corresponds to a circuit board in the receiver section as listed below.

ERROR 60 : Abnormal output signal from RF tune circuit.
ERROR 61 : Abnormal output signal from RF AMP circuit.
ERROR 62 : Abnormal output signal from IF filter circuit.
ERROR 63 : Abnormal output signal from IF AMP circuit.
ERROR 64 : LOOP 1 circuit is unlocked.
ERROR 65 : LOOP 2 circuit is unlocked.
ERROR 66 : Abnormal output signal from REF/BFO circuit.
ERROR 67 : Abnormal synthesizer control circuit or memory control circuit.

Error codes exist for the other systems tested.

This brief outline description of a modern MF/HF transceiver applies equally to other manufacturers' equipment as the technology utilized is, in most cases, very similar.

14.7   The VHF transceiver

VHF communications within the GMDSS radionet will usually be short-range on-scene communications carried out either by a hand-held unit or a fixed system on the bridge of a ship, an SAR unit or an aircraft. To satisfy GMDSS regulations three hand-held VHF radios must be carried by all ships over 500 grt. These units are constructed to withstand the hostile environmental conditions at sea and are able to communicate on the marine VHF band. Channel 16 selection is by one-touch control. Additionally, a number of preset channels can be selected including; channel 6, 13, 15, 16, 17, and 67. Other programmable channels are usually made available. In order to preserve battery life the hand-held VHF radio transmits a power output of 1 W. However, this is sufficient to communicate with VHF fixed equipment on the bridge of a ship over the short distances needed during on-scene SAR operations.

Ships sailing in any of the GMDSS radio communications areas must also be fitted with a VHF radio-telephone providing full DSC facilities. The main transceiver unit would be mounted at the point from which the ship is navigated, with remote units fitted on each bridge wing, in the chartroom and the communications centre. As an example of this equipment, the latest JRC unit from Raytheon is the JHS–31.

In addition to those channels required for operation within the GMDSS radionet, the modern VHF transceiver provides communications on all ITU channels, the American channels, Canadian channels and up to 99 private channels. In compliance with GMDSS regulations, the unit provides fully automated distress alerting using DSC. When the VHF radio is interfaced with a navigation system such as GPS, the equipment automatically provides the following data; the identity of the call vessel; nature of the distress; time; and location. As the reader should now be aware, channel 70 is the active DSC distress/safety message channel which is monitored automatically by all vessels operating within the GMDSS.

Operating a VHF radio-telephone with DSC facilities

A modern VHF system is provided with a host of communications facilities all of which can be commanded from a keypad. Many of the VHF transceiver commands have remained unchanged for decades and, consequently, will be familiar to maritime personnel, whereas some of the more complex operations need to be studied more

Fig. 14.20 VHF transceiver incorporating DSC. (Courtesy JRC/Raytheon)

carefully. The JRC JHS–31 is an excellent example of a modern maritime VHF transceiver using a dedicated keypad for inputting commands and a LCD to display information about the operational state of the system.

To appreciate how the operation of a VHF transceiver system may be commanded it is necessary to consider the functions of the operator controls.

[VOLUME] Controls the volume of audio signal from the speaker.

[SQL] Sets the squelch threshold level of operation in order to mask the inherent atmospheric noise found on the VHF bands.

[CH16] One-touch selection for the current distress channel 16.

[P] This key, along with the numerical keys, selects private channels; P01 to P99.

[DW] Used to select dual watchkeeping on a nominated channel and channel 16.

[25/1] Transmitter radiated power selection 25 W or 1 W.

[REMOTE] Enables/disables operation of the central control panel.

[SP] Speaker on/off key.

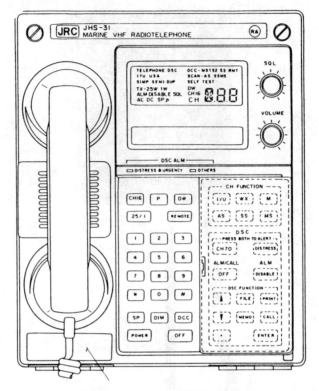

Fig. 14.21　VHF transceiver control panel. (Courtesy JRC/Raytheorn)

[DIM] Dimmer control. Four levels of control.

[OCC] Gives priority to main controller. Remote units are inhibited when this key is active.

[I/U] Selects international ITU VHF channels or USA channels.

[WX] This key, along with the numerical keys, selects weather channels in the USA and Canada. Nine receive-only channels WX1 to WX9.

[M] Ten memories M0 to M9 may be selected using this and the numerical keys.

[AS] Scan all the ITU VHF channels when active.

[SS] Use to scan operator selected channels.

[MS] Use to scan channels previously programmed into memory.

[CH70] Provides watch facility on DSC distress/safety NBDP channel 70.

[DISTRESS] Press this and [CH70] key to transmit a DSC distress call.

ALM/CALL [OFF] Used to terminate calls and alarms.

ALM [DISABLE] Used to disable audible alarms except distress/urgency calls.

[FILE] Accesses FILE mode.

[MEMO] Accesses MEMO mode.

RAPID DISTRESS ALERTING ON CH70 USING DSC		
Operation	Response	Comment
Press CH70 & DISTRESS keys simultaneously	After sending the following signal, a distress call is intermittently made at random time intervals of 3.5 to 4.5 minutes until an acknowledgement is received. Example; FORMAT : DISTRESS N of D : UNDESG DISTRESS POSITION : N54.27 W 012.45 DIST-UTC : 17.12 TELECOM : G3E SIMP TEL EOS : EOS (Display) 5T-CALL : VHF CH70 FORMAT : DISTRESS	If no navigation equipment is attached, the POSITION data are regarded as null and are transmitted as a series of 9s.

Fig. 14.22   Rapid distress alerting on CH70 using DSC. (Courtesy JRC/Raytheon)

[PRINT] Activates printer to provide output of SET-UP, MEMO and FILE contents.

[CALL] Transmits the contents of a FILE.

[.] Press to alter input data in MEMO mode and FILE mode.

[ENTER] & [*] Gives a number of command combinations.

[#] Clears mode and reset to DSC operation.

Operational parameters are clearly displayed on a LCD. A list of the less obvious display information appears below.

[OCC] Indicates which controller is being used. OCC-M = main unit, OCC-S1 remote controller S1 (S2 or S3) and OCC-RMT the central control panel is being used.

[RMT] When displayed this indicates that control from the central control panel is permitted.

Using DSC for communications

DSC operation is virtually identical to that previously described for operation within the MF/HF bands. A distress call may be sent instantly by pressing both the [CH70] [DISTRESS] keys together or, if time permits, the distress call may include further information entered via the keypad.

Whilst the DSC system follows exactly the guidelines specified by the IMO for operation within the GMDSS, it may also be used for non-distress/safety calling. As an example, figure 14.23 shows the keypad commands necessary for initializing a call to a public correspondence network requesting duplex radio-telephony connection. It is assumed that the DSC mode has been selected.

Call to Public Correspondence Network
for Request of Connection

It is presumed that DSC mode is selected by
pressing [OCC] and [ENTER].

Step No.	Operation	Response	Comments
1	[FILE] → [6] → [ENTER]	FILE 6: 000000000 FORMAT: INDIVIDUAL	
2	[·] → [▲] → [▲]	FILE 6: 000000000 FORMAT: AT/SA SERVICE	INDIVIDUAL ←┐ ↓ DISTRESS ↓ ALL SHIP ↓ AT/SA SERVICE ┘
3	[ENTER]	FILE 6: 000000000 PARTY ID: 000000000	
4	(Example): [0] → [0] → [1] [2] → [3] → [4] [5] → [6] → [7]	FILE 6: 000000000 PARTY ID: 001234567	In case of the ID-number of a coast station is 001234567.

Fig. 14.23   DSC call to a public correspondence network. (Courtesy JRC/Raytheon)

By now the reader should be aware that it is essential for the modern communications
equipment operator to be keyboard literate.

System description

The outline technical principles of VHF transceivers are similar to those of other
transceivers. Major differences do exist between VHF and MF/HF transceivers and
these are briefly listed below.

- Frequency modulation is used because the increased spectrum bandwidth required
  for this method of modulation is available on the VHF transmission band.
- The use of FM leads to the use of a pre-emphasis circuit in the TX which modifies the
  audio spectrum to assist in overcoming noise problems and improve link quality. A
  reciprocal de-emphasis circuit is provided in the receiver.
- The frequency modulator consists of a voltage controlled oscillator (VCO) and PLL
  circuit. Signal demodulation in the RX takes place in a discriminator circuit.
- Low transmitter output power requirement leads to the design of compact transceiver units.
- By definition, the use of very high transmission frequencies enables physically
  smaller antennae to be used.
- There is no requirement for manual tuning of the antenna output stage because the
  frequency spectrum bandwidth containing the whole range of communication channels is relatively small as compared with the very high frequency band used.

The technical specifications for a modern VHF transceiver are shown in figure 14.24.

Step No.	Operation	Response	Comments
5	ENTER	FILE 6: 001234567 CATEGORY: ROUTINE	
6	▲	FILE 6: 001234567 TELECOMM 1: G3E SIMP TEL	
7	· → ▲	FILE 6: 001234567 TELECOMM 1: G3E DUP TEL	DUP mode is used for telephone communication with a public correspondence network.
8	ENTER	FILE 6: 001234567 TELECOMM 2: NO INFORMATION	Only NO INFORMATION can be set.
9	▲	FILE 6: 001234567 WORK T/R:NONE	Generally, a ship station is not required to designate WORK CH.
10	▲	FILE 6: 001234567 TEL No.: 000000000	In this mode, the telephone number can be entered.
11	(Example) 8 → 1 4 → 2 → 2 4 → 5 → 9 5 → 3 → 8	FILE 6: 001234567 TEL No.: 81422459538	The preset telephone number can be called up by using · and ▲ . See MEMO 2 mode in 4.5.16.
12	ENTER	FILE 6: 001234567 EOS: ACK RQ	
13	▲	FILE 6: 001234567 FIN CALL T/R: 70	
14	CALL	Example: 24.03.90 21:45	After transmission of data in the file, the transceiver is put in standby mode.
		If an acknowledgment is given from a coast station, the designated channel and communication mode are automatically selected.	The following conditions must be established. 1. AUTO ACKNOWLEDGMENT is turned "ON" in initial setup. 2. The handset is on the hook. 3. The combination of the specified channel and communication mode is correct.

Fig. 14.23   (cont.)

VHF TRANSCEIVER OUTLINE TECHNICAL SPECIFICATIONS	
Main Unit	
TX Frequency range.	155.50 – 159.50 MHz.
RX Frequency range.	155.50 – 159.50 MHz (simplex) 160.60 – 163.50 MHz (duplex)
Channel capacity.	ITU/USA: 57. Private: 99. WX: 9. Memory: 10. Selected: 10.
Channel spacing.	25 kHz/10 kHz.
Operating modes.	Simplex/Duplex. Semi-duplex on private channels.
Types of emission.	F3E/F2B.
DSC file numbers.	15 for calling, 20 for distress call receiving, 5 for general call receiving.
Transmitter unit:	
Max. output power.	25 watts.
Reduced power.	0.1 to 1 watt.
Modulation.	FM.
Channel bandwidth.	16 kHz.
Max. freq. deviation.	$+/- 5$ kHz.
DSC specifications.	data rate: 1200 baud $+/- 30 \times 10^{-6}$. modulation: FSK. index: $2 +/- 10\%$. mark signal: 1300 Hz $+/- 10\%$. space signal: 2100 Hz $+/- 10\%$.
Receiver unit:	
System.	Double superhet.
IF.	1st: 21.4 MHz, 2nd: 455 kHz.
Sensitivity.	2 μV or less.
Selectivity.	12 kHz or more at 6 dB bandwidth.

Fig. 14.24 VHF transceiver outline technical specifications. (Courtesy JRC/Raytheon)

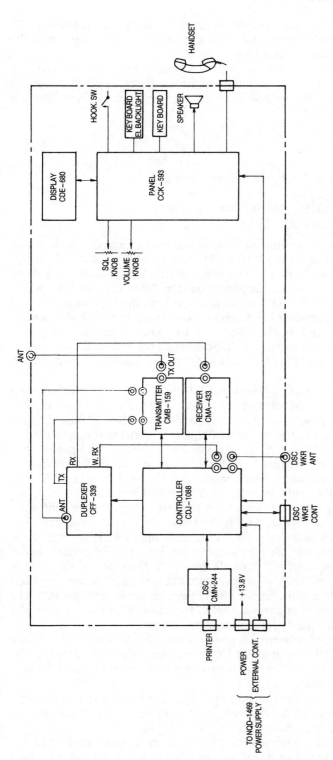

Fig. 14.25 VHF radio-telephone main unit. (Courtesy JRC/Raytheon)

A modern VHF transceiver system operates in much the same way as the MF/HF transceiver previously described. Figure 14.25 shows the overall system diagram of the JRC JHS–31 VHF transceiver fitted with DSC.

The whole system is under the command of the central microcomputer in the controller. By using a number of input/output ports the computer is able to communicate with, and command, the various circuit boards and peripherals. Data lines are available between the controller and:

- the antenna duplexer,
- the transmitter board,
- the receiver board,
- the panel circuit in the control unit (the human interface),
- the DSC printer interface,
- the separate DSC watch keeping receiver, and
- the remotely fitted control units.

System control is by executive command from the controller board, which is able actively to communicate with the TX/RX boards, the antenna duplexer, the DSC board, the printer, and the peripherals via keypad and panel board. As an example, a keypad command for channel 16 operation, via the panel board, is fed to the controller where the computer decides on what action to take. In this case both the TX and RX boards are commanded to select the frequency 156.8 MHz for simplex operation. FM baseband signal processing is activated and, in the absence of any other command, full power output from the transmitter board is selected. Indications of the actions undertaken by the computer are fed, via the panel board, to the display board where they are displayed on the LCD.

Both transmitter and receiver circuit boards hold dedicated circuitry which is commanded by the system controller. As an example, when the handset pressel is activated, speech from the microphone produces a minute voltage signal which is passed to the pre-emphasis circuit where part of the baseband frequency spectrum is emphasized in order to overcome the effect of noise in the transmission medium. The signal is then amplitude limited and bandwidth filtered to 3 kHz, in order to limit the bandwidth on transmission. Because a voice circuit has been commanded, the NOT function T-MUTE has enabled TR1 circuit. If DSC had been selected, TR1 would have been disabled and the input baseband signal would be T-DATA from the DSC circuit via the controller. The voltage controller oscillator (VCO) TR3 is the frequency modulator circuit. The VCO runs at a frequency determined by the PLL circuit appropriate for the transmission channel selected. A relatively slow variation of d.c. level derived from the audio frequency produced at the output of IC1 controls the output frequency of the VCO in sympathy with the speech signal. This frequency varying carrier signal is now amplified in IC6 and applied via further amplification in TR8 to the high power amplifier IC7. The FM channel transmitter frequency is now coupled to the antenna via the duplexer, if a duplex channel is required, or directly for simplex transmission.

Transmitter power is switched between full power (25 W) or low power (1 W) by the NOT function command line PTT input to the TX Power Controller IC9. Alternatively, an Automatic Power Control (APC) line is able to override the output from IC9 when, for instance, channel 16 is selected.

Transmitter carrier frequency is determined by the PLL which uses a feedback circuit via T-LM, the controller board and T-LA to set the channel. In this case the controller acts as the programmable divider circuit as described in section 14.5.

Referring to the receiver system diagram (figure 14.27), the received channel frequency (f_R), either via the duplexer board or directly, is coupled to the first band pass filter

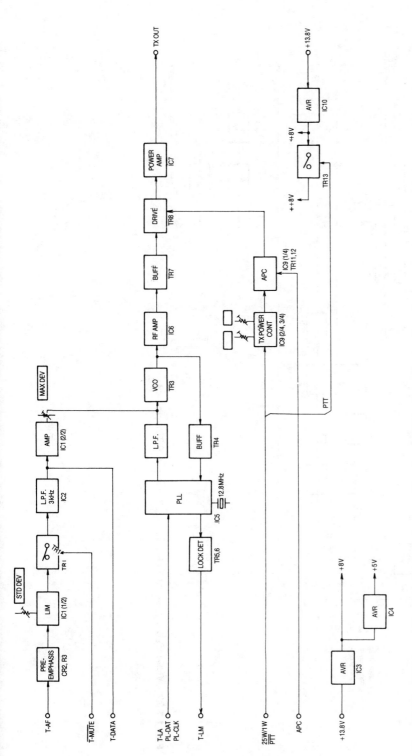

Fig. 14.26 VHF transmitter system block diagram. (Courtesy JRC/Raytheon)

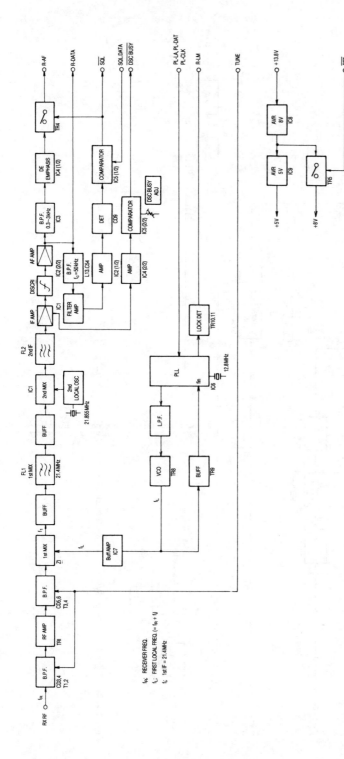

Fig. 14.27   VHF receiver system block diagram. (Courtesy JRC/Raytheon)

(BPF). The resonant frequency of this and the other BPF is selected to match the received channel by a command input TUNE. The filtered signal is now coupled to the first signal mixer where it is down converted to become the first intermediate frequency (f_i) of 21.4 MHz. A stepped variable local oscillator signal (f_L) is generated by a PLL circuit. Receiver channel selection is made via a feedback loop from the PLL, via R-LM the controller board and PL-LA, PL-DAT and PL-CLK, with the computer on the controller board acting as a frequency divider. Bandpass filter FL1 provides further signal filtering before the f_i signal is further down-converted, by the second mixer, to 455 kHz. Amplifier IC1 provides a large amount of signal gain before the IF signal is applied to the discriminator circuit which retrieves the baseband audio signal. Circuit IC3 band-pass filters the audio bandwidth to 300 Hz – 3000 Hz, after which the de-emphasis circuit modifies the audio signal spectrum in a reciprocal way to that done by the pre-emphasis circuit in the transmitter. DSC received data is coupled from the AF amplifier to the controller board via the output pin R-DATA.

System self-testing

Self-testing functions of the main unit automatically operate at pre-determined regular intervals. An error code, indicating any defective unit, is displayed for ten seconds before the LCD returns to the normal operating mode.

Error codes are displayed in three-digit form. By reference to the charts in figure 14.28 a faulty area can be diagnosed.

As an example:

E 1 3

E indicates an error display.
1 indicates a specific printed circuit board.
3 indicates the nature of the error.

As with all the equipment descriptions in this book it is suggested that the reader consults the technical manuals supplied with specific equipment. This book may be used as a reference guide to enable understanding of technical manuals.

SELF TESTING CHART – BOARDS AND ERRORS	
	Classification of PC boards.
Error display.	PC board name.
E 0 ×	Operation mistake or other error.
E 1 ×	Control.
E 2 ×	Panel.
E 3 ×	DSC.
E 4 ×	Transmitter.
E 5 ×	Receiver.
E 6 ×	Watchkeeping RX. *
E 7 ×	Central console. *
E 4A	ANTENNA. – Irregular reflection. (SWR error)
	Nature of Errors.
E 0 1	The memory channel, private channel, SS channel, or MS channel are not registered.
E × 2	EEPROM R/W error.
E × 3	RAM R/W error.
E × 4	UART communication error.
E × 5	Time-out.
E × 6	Overrun.
E × 7	Unlock.
E × 8	Modem loop test error. *
E × 9	TX/RX modem loop test error. *
* Activate the self-diagnosis function for these tests.	

Fig. 14.28   Self testing boards and errors. (Courtesy JRC/Raytheon)

15
The future

Despite the massive expansion of global satellite communications, maritime terrestrial communications will continue to be used, particularly for on-scene communications. Hand-held VHF equipment is still the most convenient method of communicating with a liferaft or an individual in the water.

However, global HF communications using the ionized layers will probably become virtually non-existent during the next decade due, particularly, to the introduction of cheap voice-grade satellite SESs. High-quality, reliable global communications are now being provided by the relatively cheap Inmarsat-M system which will undoubtedly have a massive impact on terrestrial systems. A possible saviour may be the fact that Inmarsat-M is not yet approved for use within the GMDSS, whereas the HF terrestrial communications service is approved.

Glossary of terms and abbreviations

AAIC:	Accounting authority for satellite traffic invoicing.
A/D:	Analogue-to-digital signal conversion.
ADE:	Above decks equipment. The electronic and mechanical equipment above decks on a ship.
AFC:	Automatic frequency control.
AGC:	Automatic gain control.
AHC:	Ampere hour capacity. The figure used to indicate the quantity of energy which may be delivered by a specific battery for a stated time period.
ALOHA:	A random access protocol for accessing a packet satellite network. Developed at the University of Hawaii and using the Hawaiian word for greeting.
AM:	Amplitude modulation.
AMERC:	The Association of Marine Electronic and Radio Colleges (based in Great Britain).
AORE:	Atlantic Ocean region east.
AORW:	Atlantic Ocean region west.
APC:	Adaptive predictive coding.
Apogee:	The furthest point which a satellite reaches in its orbit from the earth.
ARQ:	Automatic request repeat. An error control protocol used in data communications. The RX is able to request the retransmission of packets received in error.
ASCII:	American Standard Code for Information Exchange.
ASM1 and 2:	At-sea-maintenance qualifications, level one and two.
Azimuth:	Surface angle extended between a line of longitude and a specific direction.
Baud:	Standard unit of signalling at a rate of one signal element/second.
BCH:	Bose-Chadhuri-Hocquenghem code.
BDE:	Below decks equipment. The electronic/mechanical equipment of an SES inside the ship.
BE:	Billing entity. An accounting authority for satellite traffic invoicing.
BER:	Bit error rate. The number of bits transmitted that are received with errors, compared with the total number of bits transmitted.
BITE:	Built-in test equipment.
BPSK:	Binary-phase-shift keying. Modulation method containing two states of phase shift.

C-band:	Name of frequency band used for satellite links with a fixed station. Approximately 6/4 GHz.
CCIR:	Comite Consultatif International des Radiocommunications.
CCITT:	Comite Consultatif International Telegraphique et Telephonique.
CDMA:	Code-division multiple access.
CEPT:	Conference of European administrations.
CES:	Coast earth station.
C/N:	Carrier-to-noise ratio.
C/No:	Carrier-to-noise spectral density.
Companding:	The process of compression and expansion of the baseband signal in a communications link.
COMSAT:	Communications Satellite Corporation.
Convolution coding:	A system of encoding data bits for FEC.
COSPAS:	Cosmicheskaya Sistyeme Poiska Avariynich Sudov.
COSPAS-SARSAT:	International satellite-based emergency alerting and locating system.
CSC:	Common signalling channel.
D/A:	Digital-to-Analogue signal conversion.
dB:	Decibel.
DCE:	Data circuit terminating equipment.
Demand assignment:	A system whereby communications channels are assigned on demand from a common pool.
Demodulator:	The receiver circuitry which reproduces the baseband signal from the received signal.
DeMUX:	The method of extracting, at the receiver, multiplexed signals from a common carrier frequency.
Discriminator:	An FM demodulator circuit.
DMG:	Distress message generator.
DM:	Delta modulation.
DoR:	Data over radio.
Downlink:	Link from satellite to ground.
DSC:	Digital selective calling. An NBDP transmission used for priority alerting within the GMDSS.
DTE:	Data terminal equipment.
Duplexer:	A circuit which enables both the TX and RX stages in a transceiver to operate simultaneously.
Duplex operation:	Two way communication operation using two frequency channels permitting simultaneous transmit/receive.
EGC:	Enhanced group call. Group calling on Inmarsat-C.
EIRP:	Effective isotropically radiated power. The measure of the effectiveness in radiating transmitted power in relation to a theoretical isotropic antenna.
ELT:	Emergency locator transmitter. Airborne distress beacon in the GMDSS.
ELV:	Expendable launch vehicle.
EME:	Externally mounted equipment.
ENT:	Equivalent noise temperature.
EPIRB:	Emergency position indicating radio beacon. Maritime beacon used in the GMDSS.

Error rate:	The ratio of the incorrectly received to the correctly received bits in a data system.
ESA:	European Space Agency.
FDM:	Frequency division multiplexing.
FDMA:	Frequency-division multiple access. Method of user access to a frequency band by access to small frequency slots.
FEC:	Forward error correction.
FleetNET:	Inmarsat EGC-based system permitting shipowners to broadcast to some, or all, of their fleet.
FM:	Frequency modulation.
Forward link:	The fixed-to-mobile link.
Frequency spectrum:	The finite resource used for the propagation of radiowaves.
From-mobile:	The inbound calling direction.
GEO:	Equatorial geostationary orbit.
GES:	Ground earth station.
GF:	Gold Franc.
GMDSS:	Global maritime distress and safety system.
GRT:	Gross registered tonnes.
G/T:	Gain-to-noise temperature ratio.
Half duplex channel:	A channel which is only capable of passing information in one direction at a time.
HEO:	High elliptical orbit.
HPA:	High power amplifier.
H3E:	Amplitude modulation mode containing full carrier power and one intelligent sideband.
HSD:	High-speed data.
ID:	SES identification number. See IMN.
IFRB:	International frequency registration board. Prevents chaos by allocating frequencies in accordance with radio regulations.
IME:	Internally mounted equipment.
IMF:	International Monetary Fund.
IMN:	Inmarsat mobile number. MES identification number. Also known as the SES identification number ID.
IMO:	International Maritime Organization.
INM2:	Second generation of Inmarsat satellites.
INM3:	Third generation satellites.
Inmarsat:	International Maritime Satellite Organization.
Inmarsat-A:	Current main sat-com system for voice/data.
Inmarsat-B:	New system using all-digital technology. To replace Inmarsat-A.
Inmarsat-C:	Current small, cheaper sat-com system for data only.
Inmarsat-E:	L-band Inmarsat emergency position indicating radio beacons.
Inmarsat-M:	An all-digital cost effective sat-com system.
Intelsat:	International telecommunications satellite consortium.
Intermodulation:	Interference caused by adjacent frequency division multiplexed carriers modulating with each other.
IOR:	Indian Ocean region.
ISB:	Independent sideband operation.
ISDN:	Integrated services digital network.
ISL:	Interstation signalling link.

ISO:	International organization of standardization.
ITA:	International telegraph alphabet code.
ITU:	International Telecommunications Union.
J3E:	Amplitude modulation mode containing virtually zero carrier power and one intelligent sideband.
Lambda (λ):	The Greek character used to denote wavelength.
L-band:	Frequency band used for links between mobile and satellites. Approximately 1.5/1.6 GHz.
LEO:	Low earth orbit.
LES:	Land earth station.
Link call:	Terminology used to describe a duplex radio-telephone call.
LNA:	Low noise amplifier.
LUT:	Local user terminal. An earth station in the Cospas-Sarsat system.
Marisat:	The original maritime satellite communications system. Established in 1976 by a US consortium and taken over by Inmarsat in 1982.
MCC:	Mission control centre. Controls RCCs in Cospas-Sarsat system.
MCS:	Maritime communications sub-system on the Intelsat V generation of satellites.
MERSAR:	Merchant Ship Search and Rescue manual. Produced by the IMO.
MES:	Mobile earth station.
MID:	Maritime identification digits
MTBF:	Mean time between failures.
MMSI:	Maritime mobile service identity number.
MMSS:	Maritime mobile satellite service.
Modem:	Analogue-to-digital, and vice versa, modulator for data transmission on analogue circuits.
Modulator:	The circuit in a transmitter which mixes the baseband signal with the carrier frequency.
MRCC:	Maritime Rescue Coordination Centre.
MSI:	Maritime safety information.
MSS:	Mobile satellite services.
MTBF:	Mean time between failures. A measure of the reliability of electronic equipment.
MUF:	The maximum usable frequency.
Multiplexing:	The term used for a method whereby a number of signals are modulated onto a common carrier. Occasionally abbreviated to MUX.
NASA:	National Aeronautics and Space Administration.
NAVAREA:	IMO global navigational area.
Navtex:	NBDP broadcasting system for maritime safety information.
NBDP:	Narrow band direct printing. Narrow band transmission of teletype text.
NCC:	Network control centre.
NCS:	Network coordination station.
NOAA:	National Oceanic and Atmospheric Administration.
OCC:	Operations control centre. Inmarsat.
OTF:	The optimum traffic frequency.

O-QPSK:	Offset quaternary-phase-shift keying.
PABX:	Private automatic branch exchange.
Packet:	A self-contained component of a message which can be transferred as an entity in a data communications system. Contains the address, control and data signals.
PC:	Personal computer.
PCM:	Pulse code modulation.
PEP:	Peak envelope power.
Perigee:	The closest point to the earth of a satellite in earth orbit.
PFD:	Power flux density.
Phased array:	An antenna assembly whose directivity is controlled by signal phasing and not by mechanical means.
PLB:	Personal locator beacon (land).
PM:	Phase modulation.
Polarization:	The orientation in space of the electric field of a radiowave.
Polling:	A service whereby selected MESs are interrogated.
POR:	Pacific Ocean region.
Priority 3:	Inmarsat designation for distress calls via satellite.
Protocol:	A discipline controlling the operation of a system.
PSDN:	Public switched data network.
PSPDN:	Public switched packet data network.
PSTN:	Public switched telephone network.
PSK:	Phase-shift keying. The generic term for a method of modulation used in digital communications systems. See also BPSK, QPSK, O-QPSK.
PVT:	Performance verification test. An automatic test performed by an MES to verify that it is functioning correctly.
QPSK:	Quaternary-phase-shift keying. Modulation method permitting double the data rate possible with BPSK.
Radome:	The protective cover of an SES ADE system.
RCC:	Rescue Coordination Centre.
Return link:	The mobile-to-ground link.
R3E:	Amplitude modulation mode containing reduced carrier power and one intelligent sideband.
RX:	Receiver.
SafetyNET:	Inmarsat EGC-based system for MSI.
SAR:	Search and rescue.
SARSAT:	Search and rescue satellite-aided tracking.
SART:	Search and rescue radar transponder. Radar beacon which indicates its position in response to surface or airborne radar signals.
SAW:	Surface acoustic wave bandpass signal filter.
SCC:	Satellite Control Centre.
SCPC:	Single channel per carrier.
SDR:	Special drawing rate.
Semi-duplex operation:	Two-way communication using a single frequency channel but not permitting simultaneous transmit/receive.
SES:	Ship earth station.
SFU:	Store and forward unit.
Sideband:	The resultant upper and lower frequency bands produced either side of a carrier frequency as result of modulation.

Sidelobes:	Unwanted radiation pattern of an antenna. RF energy in the sidelobe is lost from the main lobe and is wasted.
Simplex Operation:	Two-way communications operation using a single frequency channel.
SOLAS:	Safety of Life at Sea Convention.
Spot Beam:	Shaped satellite transmission beam designed to cover a specific area of the earth's surface.
SSFC:	Sequential single-frequency code system. International selective calling system in accordance with Appendix 39 of the ITU radio regulations. To be replaced by the DSC in the GMDSS.
Step track:	The method used to move an MES antenna while searching for a satellite lock.
SWR:	Standing wave ratio. An indication of the quality of the antenna and feeder system.
TDM:	Time division multiplexing.
TDMA:	Time-division multiple access. Multiple access technique using time slots.
To-mobile:	The outbound calling direction.
TOR:	Telex over radio.
Traffic:	Information passed over a communications link.
Transponder:	Satellite communications unit comprising, receive antenna, receiver, filters, amplifiers, output filters and transmit antenna.
TT&C:	Tracking, telemetry and command system. Ground control system to monitor and control remotely a satellite.
TWTA:	Travelling-wave tube amplifier. A microwave amplification device used in a satellite.
TX:	Transmitter.
UTC:	Coordinated universal time.
Uplink:	The link from ground to satellite.
UW:	Unique word.
VDU:	Visual display unit.
VOC:	Voice operated carrier.
VSWR:	Voltage standing wave ratio.
WARC:	World Administrative Radio Conference. Sub group of ITU to produce the regulations governing the use of radio frequencies.
WMO:	World Meteorological Organization.
WX:	The abbreviation for weather messages.
X25 and X400:	Standards for messaging and data systems specified by the CCITT.

Appendix 1
EXAMINATION SYLLABUS
GUIDELINES FOR THE GMDSS
GOC

A. Knowledge of the MMS and the MMSS

A.1. General principles and basic features of the MMS.

1. Types of communication in the MMS;
   - distress, urgency and safety; public correspondence; port operations; ship movement; intership and on-board communications.
2. Types of ship station in the MMS;
   - ship; coast; pilot; port; aircraft stations and maritime rescue coordination centre (MRCC).
3. Elementary knowedge of frequencies and bands;
   - concept of frequency; equivalence between frequency and wavelength; units of frequency; subdivisions of radio frequency spectrum.
4. Characteristics of frequencies;
   - propagation in free space; ground wave; ionospheric propagation; propagation of MF, HF, VHF and UHF.
5. Knowledge of the role of the various modes of communications;
   - DSC; R/T; NBDP; fax; data and telegraphy.
6. Elementary knowledge of different types of modulation and classes of emission;
   - classes of emission; carrier and assigned frequencies; bandwidth of different emissions; official designation of emissions (H3E etc) and unofficial designations (TLX etc)
7. Frequencies allocated to the MMS.
   - usage of frequencies, MF,HF,VHF,UHF,SHF, in the MMS; concept of the radio channel, simplex, duplex etc; frequency plans and channelling systems for HF,VHF,HF,NBDP with reference to the relevant appendices of the Radio Regulations; GMDSS distress and safety frequencies; non-GMDSS distress and safety frequencies; calling frequencies.

A.2. General principles and basic features of the MMSS.

1. Maritime satellite communications.
   - Inmarsat space segment; modes of communication—telex, telephone, data, fax, store and forward operation; distress communications; Inmarsat-A services; Inmarsat-C services; EGC service.
2. Types of station in the MMSS.
   - CES; NCS; SES.

B.   Detailed practical knowledge and ability to use the basic equipment of a ship station

B.1.   Use in practice the basic equipment of a ship station.

1. Watchkeeping receivers.
   - the controls and usage of the following watch receivers; 2182 kHz; VHF DSC; MF DSC; MF/HF DSC.
2. VHF radio installation:
   - channels; controls; usage; DSC.
3. MF/HF radio installation:
   - frequencies; typical controls and usage e.g. connecting power; selecting RX/TX frequency, class of emission and ITU channel number; tuning the TX; using volume control, squelch, clarifier, RX fine tuning, AGC, 2182 kHz instant selector, alarm generator; controlling RF gain; testing the alarm generator.
4. Antennae:
   - isolators; VHF whip; MF/HF whip; MF/HF wire antennas; construction of an MF emergency antenna.
5. Batteries:
   - different kinds and characteristics; charging; battery maintenance.
6. Survival craft radio equipment.
   - portable two-way VHF R/T; SART; EPIRB.

B.2.   Digital Selective Calling.

1. Call format specifier:
   - calls—distress; all ships; group; individual; geographic area; automatic/semi-automatic.
2. Call address selection with the MMSI number system:
   - nationality identifier; group calling numbers; coast station numbers; MMSI number with three trailing zeros.
3. Call categorization:
   - distress; urgency; safety; ship business; routine.
4. Call telecommand and traffic information:
   - distress alerts; other calls; working frequency information.

B.3.   General principles of NBDP and TOR systems. Use maritime NBDP and TOR equipment in practice.

1. NBDP systems:
   - automatic; semi-automatic; manual, ARQ mode; FEC mode; ISS/IRS arrangement; master/slave; radio telex number; answerback; numbering of the SSFC calling system.
2. TOR equipment:
   - controls and indicators; keyboard operation.

B.4.   Usage of the Inmarsat system. Use Inmarsat equipment or simulator in practice.

1. Inmarsat-A SES:
   - satellite acquisition; telex and telephone services; data and fax communications.
2. Inmarsat EGC RX:

- pre-programming an SES for EGC message reception; selecting operating mode for EGC reception.
3. Inmarsat-C SES:
  - components of an Inmarsat-C SES; entering/updating position; usage of an Inmarsat-C SES; sending and receiving test messages.

B.5. Fault locating.

1. Proficiency in elementary fault locating by means of built-in measuring instruments or software in accordance with equipment manuals. Elementary fault repair such as replacement of fuses, indicator lamps etc.

C. Operational procedures and detailed practical operation of GMDSS systems and subsystems

C.1. The GMDSS.

1. Sea areas and GMDSS master plan.
2. Watchkeeping on distress frequencies.
3. Functional requirements of ship stations.
4. Carriage requirements of ship stations.
5. Sources of energy of ship stations.
6. Means of ensuring the functionality of ship station equipment.
7. Licences; radio safety certificates; inspections and surveys.

C.2. Inmarsat usage in the GMDSS.

1. Inmarsat-A SES:
  - distress communications and services; use of the distress facility; satellite acquisition; telex and telephony distress calls; procedures for distress calls; RCCs associated with CESs.
2. Inmarsat-C SES:
  - distress communications and services; sending a distress alert; sending a distress priority message; 2-digit code sequence.
3. Inmarsat EGC:
  - purpose of the EGC service; all-ships messages and Inmarsat system messages; classes of Inmarsat-C SES and their EGC reception.

C.3. NAVTEX.

1. The NAVTEX system:
  - purpose; frequencies; reception range; message format.
2. The NAVTEX receiver:
  - selection of transmitters; message type; message which cannot be rejected; use of subsidiary controls; changing paper.

C.4. EPIRBs.

1. Satellite EPIRBs:
  - basic characteristics of operation of 406 MHz EPIRB; 121.5 MHz EPIRB including homing function; basic characteristics of operation on 1.6 GHz, information

content of a distress alert; manual usage; float-free function; routine maintenance including—testing; checking battery expiry date; cleaning float-free release mechanism.
2. VHF-DSC-EPIRB:
   ● as above.

C.5.   SART.

1. SART:
   ● technical characteristics; operation; range of a SART TX; routine maintenance including checking battery expiry date.

C.6.   Distress, urgency and safety communications procedures in the GMDSS.

1. Distress communications.
   DSC distress alert:
   ● definition of a distress alert; transmission of a distress alert; shore-to-ship distress alert relay; distress alert by station not itself in distress.

   Reception and acknowledgement of a DSC distress alert:
   ● acknowledgement procedure by telephony and DBDP; receipt and acknowledgement by coast and ship stations.

   Handling of distress alerts:
   ● preparations for handling of distress traffic; distress traffic terminology; testing DSC distress and safety calls; on-scene communications; SAR operation.
2. Urgency and safety communications:
   ● meaning of urgency and safety; procedures for DSC urgency and safety calls; urgency communications; medical transports; safety communications.
3. Communications by radio-telephony with stations of the old distress and safety system:
   ● radio-telephone alarm signal; distress signal, call and message; acknowledgement of distress message; distress traffic terminology; transmission of a distress message by a station not itself in distress; medical advice.
4. Reception of MSI:
   ● reception by NAVTEX, Inmarsat EGC, HF NBDP; navigational warning system of the old distress and safety system; navigational warnings transmitted by radio-telephony.
5. Protection of distress frequencies:
   ● guard bands; tests of distress frequencies; transmissions during distress traffic; avoiding harmful interference; prevention of unauthorized transmissions.

C.7.   SAR.

1. The role of RCCs.
2. Merchant Ship Search and Rescue manual: MERSAR.
3. Maritime rescue organizations.
4. Ship routeing systems.

D. Miscellaneous skills and operational procedures for general communications.

D.1. Ability to use English language, both written and spoken, for the satisfactory exchange of communications relevant to the safety of life at sea.

1. Use of the International Code of Signals and the IMO Standard Maritime Navigational Vocabulary/Seaspeak.
2. Recognizes standard abbreviations and commonly used service codes.
3. Use of the international phonetic alphabet.

D.2. Obligatory procedures and practices.

1. Effective use of obligatory documents and publications.
2. Radio record keeping.
3. Knowledge of the regulations and agreements governing the MMS and the MMSS.

D.3. Practical and theoretical knowledge of general communication procedures.

1. Selection of appropriate communication methods in different situations.
2. Traffic lists.
3. Radio telephone call:
   - methods of calling a coast station; ordering a manually switched link call; ending the call; special facilities of calls; method of calling a coast station using DSC; selecting an automatic radio-telephone call.
4. Radio telegram.
   The parts of a radio telegram:
   - preamble; service instructions and indications; address; text; signature.
     Addresses:
   - full; registered; telephonic; telex.
   Counting of words.
   Transmission by radio-telephony/radio-telex.
5. Traffic charges:
   - international charging system; Inmarsat communication charging system; AAIC; landline, coast charge and ship charge; currencies used in international charging.
6. Practical traffic routes.
7. World geography, especially the principal shipping routes and related communication traffic routes.

Appendix 2
THE DECIBEL (dB)

Extremely large ranges of signal parameters are used in electronic communications work. For instance, power may be quoted in values as small as microwatts or as large as megawatts. Ranges as large as this also apply to voltage, current, frequency and, to a lesser extent, temperature.

A *bel* (whole unit) is a ratio of two different levels. The example is an amplifier whose signal power-out level over the power-in signal level gives the gain figure as a ratio. This figure is often quoted in dB.

The bel unit was named in honour of Alexander Graham Bell, famous for his research work leading to the invention of the telephone. However, for most electronic work the bel as a unit is much too large and, consequently, a one tenth unit, the decibel is used.

The decibel is not a linear unit of measurement. This is primarily because it would be virtually impossible to find a linear unit which was readily available to show very large changes of signal parameters.

It should be noted that the decibel is based on ratio measurement and can have no absolute value unless it is referenced to a specific level. As an example the figure 20 dB above 1 mW possesses an absolute value since a reference level is quoted. From the table, at the end of this appendix, it can be seen that 20 dB possesses a power ratio of 100. With respect to 0 dB there will be a positive increase of 100 and a negative decrease of the same figure (0.01).

A figure often quoted to indicate power levels is 3 dB. This corresponds to a doubling of the gain power ratio from 1.0 to 2.00 or a halving of the loss power ratio from 1.0 to 0.50. Thus, the −3 dB points of a bandwidth graph are often called the 'half power points'. The gain power ratio is doubled for each 3 dB rise and halved for each −3 dB change.

Decibel figures are calculated using logarithmic tables or electronic calculators. It is not always convenient to do this, and if an approximate ratio is required reference can be made to a table of decibels.

Once the various signal parameters have been converted to the common ratio—decibels—the total figure for gain or loss in a system may be found by simply adding and/or subtracting the dB ratios.

Common reference levels used in communications systems

The dBm: The reference level used is the milliwatt (mW). The m signifies that the dB figure is quoted to a reference level of 1 mW. Therefore dB values will represent an absolute power level.

The dBW: Reference level: the Watt. The W signifies that the dB figure is quoted as a power ratio to a reference level of 1 W.

The dBμV: Reference level: the μVolt. The μV signifies that the dB figure is quoted as a voltage ratio to a reference level of 1 μV.

The dBi: Reference level: the mAmp. The i signifies that the dB figure is quoted as a current ratio to a reference level of 1 mA.

The dBHz: Reference level: the Hertz. The Hz signifies that the dB figure is quoted as a frequency ratio to a reference level of 1 Hz.

The dBK: Reference level: the Kelvin. The K signifies that the dB figure is quoted as a temperature ratio to a reference level of 1 K.

Decibel ratio				
	Gain (+)		Loss (−)	
dB	Power ratio	Voltage ratio	Power ratio	Voltage ratio
0	1.00	1.00	1.00	1.00
1	1.26	1.12	0.80	0.89
2	1.59	1.26	0.63	0.79
3	2.00	1.41	0.50	0.71
4	2.51	1.59	0.40	0.63
5	3.16	1.78	0.32	0.56
6	3.98	2.00	0.25	0.50
7	5.00	2.24	0.20	0.45
8	6.31	2.51	0.16	0.40
9	7.96	2.82	0.12	3.36
10	10.00	3.16	0.10	0.32
15	31.60	5.62	0.03	0.18
20	100.00	10.00	0.01	0.10
50	10^5	316.20	10^{-5}	3.16×10^{-3}
80	10^8	10^4	10^{-8}	1.00×10^{-4}
100	10^{10}	10^5	10^{-10}	1.00×10^{-5}

Appendix 3
PRODUCERS OF SHIP TERMINALS

ABB NERA A.S., (Norway) Mr Ottar B Bjastad, Tel: (47) 2 844-700 Fax: (47) 2 982-267
Inmarsat-A Maritime, Inmarsat-A Land Transportable, Inmarsat-C

ANRITSU, (Japan) Mr K Shimizu, Radio Products Divn, Tel: (81) 3 3446-1111 Fax: (81) 3 3440-2430
Inmarsat-A Maritime, Inmarsat-C

CLERITEK, (USA) Mr Bob Jones, VP Sales & Marketing, Tel: (1) 408 433-0335 Fax: (1) 408 433-0991
Inmarsat-C

DMT, (Germany) Mr Alois Salfner, Tel: (49) 4088 252-845 Fax: (49) 4088 254-105
Inmarsat-A Maritime, Inmarsat-C

DORNIER, (Germany) Mr Friedrich Schmidt, Tel: (49) 7545 84253 Fax: (49) 7545 84411
Inmarsat-A Maritime

FURUNO, (Japan) Mr Hiroaki Komatsu, Gen Mgr Intl Marketing Divn, Tel: (81) 798 652-111 Fax: (81) 798 654-200
Inmarsat-C

HAGENUK, (Germany) Mr Nils Schwarz, Tel: (431) 8818-4242 Fax: (431) 8818-4368
Inmarsat-C

JAPAN RADIO COMPANY, (Japan) Mr Akeyoshi Sumita, Exec MD, Tel: (81) 3 3584-8756 Fax: (81) 3 3584-8891
Inmarsat-A Maritime, Inmarsat-A Land Transportable, Inmarsat-C

KELVIN HUGHES, (UK) Mr John Almond, Tel: (44) 81 500-1021 Fax: (44) 81 500-0837
Inmarsat-C

KODEN ELECTRONICS CO LTD, (Japan) Mr Hidekazu Shinkawa Tel: (81) 3 3440-3864 Fax: (81) 3 3473-3837
Inmarsat-C

MAGNAVOX, (USA) Dr Peter Williams, Satcom Product Mgr, Tel: (1) 310 618-1200 Fax: (1) 310 618-7001
Inmarsat-A Maritime
Mr Mario Cid Fernandez, Tel: (1) 516 667-7710 Fax: (1) 516 667-2235
Inmarsat-A Land Transportable

MAN, (Germany) Mr Armin Keller, Tel: (49) 8131 907-354 Fax: (49) 8131 907-372
Inmarsat-C

MARCONI MARINE, (UK) Mr Roger Branscomb, Tel: (44) 245 353-221 Fax: (44) 245 358-776
Inmarsat-A Maritime, Inmarsat-A Land Transportable

MOBILE TELESYSTEMS INC, (USA) Mr Shafiq A Chaudhuri, Tel: (1) 301 590-8500 Fax: (1) 301 590-8558
Inmarsat-A Maritime, Inmarsat-A Land Transportable

MUSSON, (Ukraine) Mr I. Kalyuzhniy, Tel: (7) 069 233-490 Fax: (7) 069 238-081
Inmarsat-A Maritime

PHILIPS RADIO, (Denmark) Mr Jens Bjarnoe Thostrup, Tel: (45) 3288-3793 Fax: (45) 3288-3930
Inmarsat-C

RADAR DEVICES INC, (USA) Mr Lawrence F Anderson, Tel: (1) 415 483-1953 Fax: (1) 415 351-7413
Inmarsat-A Maritime, Inmarsat-A Land Transportable

SAIT MARINE, (Belgium) Mr A Godts, Tel: (32) 3320-1711 Fax: (32) 3321-5034
Inmarsat-C

SCIENTIFIC ATLANTA, (USA) Mr Macy Summers, Tel: (1) 510 483-1953 Fax: (1) 510 483-1953 Fax: (1) 510 351-7413
Inmarsat-C

SHIJIAZHUANG, (Peoples Republic of China) Mr Luo Zheng Bin, Tel: (86) 311 33330 Cable: 3001
Inmarsat-A Maritime, Inmarsat-A Land Transportable

SNEC, (France) Mr R Dickens, Tel: (33) 3180-7122 Fax: (33)3180–6549
Inmarsat-C

SP RADIO, (Denmark) Mr K E Dantoft, Tel: (45) 9818-0999 Fax: (45) 9818-6717
Inmarsat-C

SPERRY, (USA) Mr George X Tsirimokos, Tel: (1) 804-974-2566 Fax: (1) 804 974-2259
Inmarsat-A Maritime, Inmarsat-C

STC INTERNATIONAL MARINE, (UK) Mr Rodger Perks, Tel: (44) 81 640-34001 Fax: (44) 81 685-0321
Inmarsat-A Maritime, Inmarsat-A Land Transportable, Inmarsat-C

THRANE & THRANE, (Denmark) Mr Per Thrane, Tel: (45) 3156-4111 Fax: (45) 3156-2140
Inmarsat-C

TOSHIBA, (Japan) Mr Mahiro Tada, Tel: (81) 3 3457-3120 Fax: (81) 3 3456-1699
Inmarsat-A Maritime
Mr Shuiji Funo, Toshiba Corp
Inmarsat-C

TRIMBLE NAVIGATION, (USA) Mr Don Green, Tel: (1) 408 481-6834 Fax: (1) 408 737-6057
Inmarsat-C

Appendix 4
SATELLITE TELEPHONE COUNTRY CODES

Afghanistan	93	Costa Rica	506
Albania	355	Cuba	53
Algeria	213	Curacao	5999
Andorra	33628	Cyprus	357
Angola	244	Czechoslovakia	42
Anguilla	1809497	Denmark	45
Antigua	1809	Djibouti	253
Argentina	54	Dominica	18090449
Aruba	297	Dominican Rep.	1809
Ascension Is.	247	Ecuador	593
Australia	61	Egypt AR	20
Austria	43	El Salvador	503
Azores	351	Equatorial Guinea	240
Bahamas	1809	Ethiopia	251
Bahrain	973	Falkland Is.	500
Bangladesh	880	Faeroe Is.	298
Barbados	1809	Fiji	679
Belgium	32	Finland	358
Belize	501	France	33
Benin	229	French Guiana	594
Bermuda	18092	French Polynesia	689
Bolivia	591	Gabon	241
Bonaire	5997	Gambia	220
Botswana	267	Germany (was DDR)	37
Bourkina Fasso	226	Germany (was FDR)	49
Brazil	55	Ghana	233
Brunei	673	Gibraltar	350
Bulgaria	359	Greece	30
Burma	95	Greenland	299
Burundi	257	Grenada	1809440
Cameroon	237	Guadeloupe	590
Canada	1	Guam	671
Canary Is.	34	Guatemala	502
Cape Verde Is.	238	Guinea	224
Cayman Is.	180994	Guinea Bissau	245
Central African Republic	236	Guyana	592
Chad	235	Haiti	509
Chile	56	Honduras	504
China	86	Hong Kong	852
Christmas Is.	6724	Hungary	36
Cocos Keeling Is.	6722	Iceland	354
Colombia	57	India	91
Congo PR	242	Indonesia	62
Cook Is.	682	Inmarsat AOR–E	871

Inmarsat POR	872
Inmarsat IOR	873
Inmarsat AOR–W	874
Iran	98
Iraq	964
Ireland	353
Israel	972
Italy	39
Ivory Coast	225
Jamaica	1809
Japan	81
Jordan	962
Kenya	254
Kiribati	686
Korea PDR	850
Korea Rep.	82
Kuwait	965
Lebanon	961
Lesotho	266
Liberia	231
Libya	218
Liechtenstein	4175
Luxembourg	352
Macao	853
Madagascar	261
Madeira	35191
Malawi	265
Malaysia	60
Maldive Is.	960
Mali	223
Malta	356
Marshall Is.	692
Martinique	596
Mauritania	222
Mauritius	230
Mexico	52
Micronesia	691
Monaco	3393
Montserrat	1809
Morocco	212
Mozambique	258
Myanmar	95
Namibia	264
Nauru Is.	674
Nepal	977
Netherlands	31
Netherlands Antilles	599
New Caledonia	687
New Zealand	64
Nicaragua	505
Niger	227
Nigeria	234
Niue Is.	683
Norfolk Is.	672
Norway	47
Oman	968
Pakistan	92
Panama	507
Papua New Guinea	675
Paraguay	595
Peru	51
Philippines	63
Poland	48
Portugal	351
Puerto Rico	1809
Qatar	974
Reunion	262
Romania	40
Rwanda	250
St Helena	290
St Kitts-Nevis	1809
St Lucia	1809
St Pierre/Miquelon	508
St Vincent/Grenadines	1809
Samoa (US)	684
Samoa (Western)	685
San Marino	39549
Sao Tome/Principe	239
Saudi Arabia	966
Senegal	221
Seychelles	248
Sierra Leone	232
Singapore	65
Solomon Is.	677
Somalia DR	252
South Africa	27
Spain	34
Sri Lanka	94
Surinam	597
Swaziland	268
Sweden	46
Switzerland	41
Syria	963
Taiwan	886
Tanzania	255
Thailand	66
Togo	228
Tonga	676
Trinidad & Tobago	1809
Tunisia	216
Turkey	90
Turks & Caicos Is.	1809
Uganda	256
UAE	971
UK	44
Uruguay	598
USA	1
USSR	7
Vanuatu	678
Vatican City	3966982
Venezuela	58
Vietnam	84

Appendix 5
SATELLITE TELEPHONE SERVICES

Some or all of the following services may be offered by coast earth stations operating in the Inmarsat-A network.

Code	Service	Remarks
00	Automatic	Use this code to make automatic telephone, facsimile and voice band data calls using International Direct Dial (IDD) codes.
11	International operator	Use this code to obtain information from the international operator of the country within which the coast earth station is situated.
12	International information	Use this code to obtain information about subscribers located in countries other than that in which the coast earth station is situated.
13	National operator	Use this code to obtain assistance to connect to subscribers within the country in which the coast earth station is situated. In some countries which do not have an international operator, use this code instead of code 11.
14	National information	Use this code to obtain information about subscribers located in the country in which the coast earth station is located.
17	Telephone call booking	This code may be used via some coast earth stations to book telephone calls, although normally this code is used via the telex service.
20	Access to a maritime PAD	This code is used when using a voice band data modem to access a Maritime Packet assembly/disassembly (PAD) facility in a packet switched public data network. The PAD is accessed via telephone circuits and the prefix 20 should be followed by two additional digits indicating the required data rate.
23	Abbreviated dialling (short code selection)	This code is used by some coast earth stations to allow Inmarsat-A equipped subscribers to utilize abbreviated dialling codes for their regularly dialled numbers.

Code	Service	Remarks
31	Maritime enquiries	This code may be used for special enquiries such as ship location, authorization etc, etc.
32	Medical advice	Use this code to obtain medical advice. Some coast earth stations have direct connections with local hospitals when this code is used.
33	Technical assistance	Use this code if you are having technical problems with your Inmarsat-A terminal. Technical staff at coast earth stations will normally be able to assist you.
34	Person-to-person call	Use this code to contact the operator for a person-to-person call.
35	Collect call	Use this code to contact the operator for a collect call (charges payable by the recipient of the call).
36	Credit card call	Use this code to charge a telephone call to a credit or charge card.
37	Time and charges	This code should be dialled at the start of a call before the code 00 for an automatic call. This service will enable the time and charges for the call being set up to be advised to the Inmarsat-A terminal operator. This will be either by a call back from the coast earth station operator on the telephone or, more normally, a short telex message containing the duration and call charges relating to the call.
38	Medical assistance	This code should be used if the condition of an ill or injured person on board the vessel requires urgent evacuation ashore or the services of a doctor aboard the vessel. This code will ensure that the call is routed to the appropriate agency/authority ashore to deal with the situation.
39	Maritime assistance	This code should be used to obtain maritime assistance if the vessel requires assistance, tow, oil pollution etc, etc.

Index